The Seven Spaces of Creation: Octonion Cosmology

Stephen Blaha Ph. D.
Blaha Research

The Spectrum of Octonion Spaces
Octonion Megaverses
Octonion Universes
SpaceTimeLess Octonion Spaces
Minispaces
A 10 Dimension Octonion Base Space for Megaverses
Three Additional Spaces for Blaha's Monads
Apparently the Deepest Level of Physical Reality
A Detailed Interrelation of the Spectrum of Spaces

Pingree-Hill Publishing
MMXX

ISBN: 978-1-7356795-3-2

Rev. 00/00/01 November 9, 2020

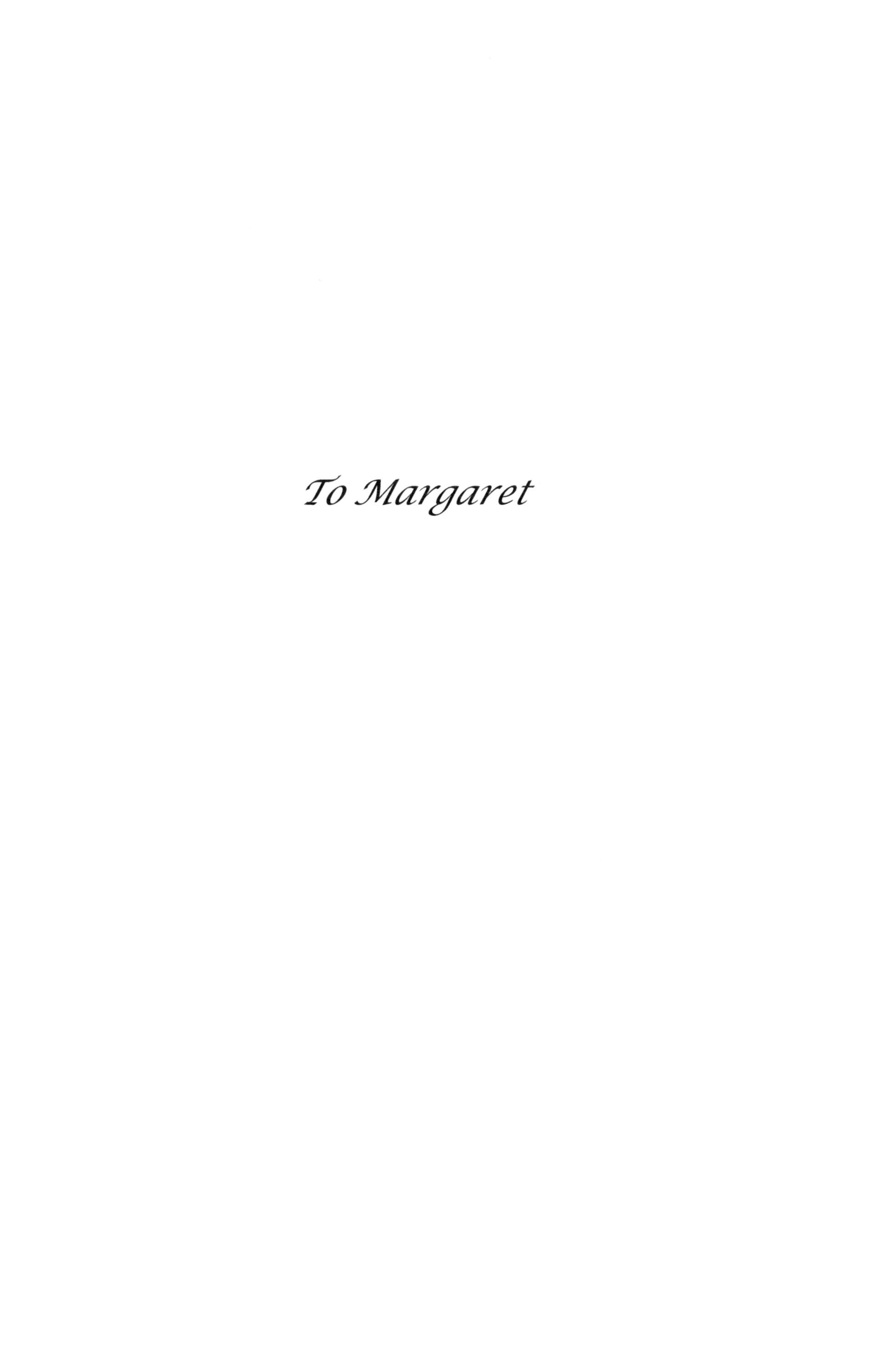

To Margaret

Some Other Books by Stephen Blaha

All the Megaverse! Starships Exploring the Endless Universes of the Cosmos using the Baryonic Force (Blaha Research, Auburn, NH, 2014)

SuperCivilizations: Civilizations as Superorganisms (McMann-Fisher Publishing, Auburn, NH, 2010)

All the Universe! Faster Than Light Tachyon Quark Starships & Particle Accelerators with the LHC as a Prototype Starship Drive Scientific Edition (Pingree-Hill Publishing, Auburn, NH, 2011).

Unification of God Theory and Unified SuperStandard Model THIRD EDITION (Pingree Hill Publishing, Auburn, NH, 2018).

The Exact QED Calculation of the Fine Structure Constant Implies ALL 4D Universes have the Same Physics/Life Prospects (Pingree Hill Publishing, Auburn, NH, 2019).

Unified SuperStandard Theory and the SuperUniverse Model: The Foundation of Science (Pingree Hill Publishing, Auburn, NH, 2018).

Quaternion Unified SuperStandard Theory (The QUeST) and Megaverse Octonion SuperStandard Theory (MOST) (Pingree Hill Publishing, Auburn, NH, 2020).

Unified SuperStandard Theories for Quaternion Universes & The Octonion Megaverse (Pingree Hill Publishing, Auburn, NH, 2020).

The Essence of Eternity: Quaternion & Octonion SuperStandard Theories (Pingree Hill Publishing, Auburn, NH, 2020).

A Very Conscious Universe (Pingree Hill Publishing, Auburn, NH, 2020).

Hypercomplex Universe (Pingree Hill Publishing, Auburn, NH, 2020).

Why is the Universe Real? From Quaternion & Octonion to Real Coordinates

Available on Amazon.com, bn.com Amazon.co.uk and other international web sites as well as at better bookstores (through Ingram Distributors).

CONTENTS

FIGURES and TABLES

Introduction

This volume presents a complete Cosmology based on a spectrum of seven octonion spaces and three spaces of functionals. It is a truly fundamental physical theory that “explains” space-time, internal symmetry groups, and elementary particles, and their associated detailed phenomena. It provides a justification for the Unified SuperStandard Theory that the author derived logically over the past twenty years based on the author’s extension of Quantum Field Theory presented in papers in the 1970s.

Thus one sees a panorama ranging from the most fundamental physical levels to current elementary particle physics experiments. The book, *Cosmic Seeds of Universes and Megaverses*, is presented here for completeness and the reader’s convenience.

1. The Spectrum of Octonion Spaces

Creation mirrors the Creator.

In earlier books in 2020 (See References) we successfully based the Unified SuperStandard Theory (UST) on the Quaternion Unified SuperStandard Theory (QUeST) with a remarkable match between the internal symmetries and space-time symmetry of both theories.

We now describe a Cosmology based on a spectrum of octonion spaces that includes interlocked universes and Megaverses. This Cosmology appears to be the deepest level of physical reality. In this chapter we outline the features of its seven spaces. We describe the details of the spaces and their interrelations in the chapters that follow.

1.1 Octonion Basis

The most general type of number that can support quantum field theory (and perturbation theory) are octonions. Octonions have significant properties that enable them to be used in a quantum field theory development:

1. An octonion is an 8-tuple of real numbers. A complex octonion is an 8-tuple of complex numbers.
2. Octonions are non-associative.
3. Octonions are one of the two finite dimensional division rings having the real numbers as a proper subring. (The other is quaternions—considered in chapter 5.)
4. Octonions are non-commutative. (This is not a roadblock for quantum field Theory, which is also non-commutative in general.)

We can represent a complex octonion (bioctonion) b as

$$b = b_{real} + Ib_{imaginary}$$

where b_{real} and $b_{imaginary}$ are real-valued octonions, and I is an additional fundamental octonion unit . We can also represent a complex octonion as an octonion with complex coordinates.

The octonion spaces in this book are based on treating octonions as the source of dimensions. Octonion unit algebra is not used.

1.2 The Spectrum of Physical Octonion Spaces

There is a set of seven conceptually interrelated spaces built on octonions. These spaces are listed in Fig. 1.1. Further types of octonion spaces can be defined. But they do not appear to be necessary, and their connection to the seven spaces can be eliminated (truncated) by the specifications of the Octonion Octonion Octonion[1] space and the Quaternion Minispace. We describe the limitation to seven spaces in later chapters.

Coordinates	Number of Coordinates	Dimension Array Size	Space-Time Dimensions
Spaceless[2]			
Octonion Octonion Octonion (512)	Octonion Octonion Octonion	512×512	0
Quaternion Octonion Octonion (256)	Quaternion Octonion Octonion	256×256	0
Complex Octonion Octonion (128)	Complex Octonion Octonion	128×128	0
Cosmology[3]			
Octonion Octonion (64)	Octonion Octonion	64×64	10 quaternion octonion
Quaternion Octonion[4] (32)	Quaternion Octonion	32×32	8 complex octonion
Complex Octonion[5] (16)	Complex Octonion	16×16	4 octonion
Minispace[6]			
Quaternion (4)	Quaternion	4×4	0

Figure 1.1. The spectrum of octonion spaces. The numbers in parentheses in column 1 are the number of dimensions in the respective coordinate. The items in column 2 are the number of rows of dimensions (512, 256, 128, 64, 32, 16, 4).

1.3 Characteristics and Roles of the Seven Spaces

The seven spaces of Octonion Cosmology have a variety of features that ultimately lead to the Unified SuperStandard Theory (UST). We describe the relationships in chapter 7. UST bears direct comparison to experiment and accommodates the known features of The Standard Model of Elementary Particles.

The spectrum of seven spaces is also of interest in its own right.[7]

---

[1] This phrase, and similar phrases, are not typos. The phrase "Octonion Octonion Octonion" means 8 octonion octonions, which means 64 octonions, which means a 512-tvector for a coordinate. The coordinate has 512 real values assembled in a 512-vector. Similar comments apply to other entries in Fig. 1.1.

[2] Spaceless spaces are those without space-times. Thus they are "point" spaces.

[3] Cosmological spaces are those, which are directly related to physics from universes to elementary particles.

[4] In our earlier books in 2020 we also designated this 1024 dimension space as 64 complex octonion space.

[5] In our earlier books in 2020 we also designated this 256 dimension space as 32 complex quaternion space.

[6] A Minispace is a space (universe) with no space-time that can be embedded in QUeST space-time.

[7] We explore these spaces from a physical perspective. It is also possible to view these spaces as analogues of Philosophical-Religious phenomena. For example, the three spaceless spaces can be viewed as an analogue of a deity; and the 64 dimension Octonion Octonion space with 10 dimensions and 32-spinor fermions could be viewed as an analogue of the concept of 10 Sefirot with 32 emanations in Judaism. Their qualitative features are remarkably

1.3.1 Spaceless Spaces

The three spaceless spaces of Fig. 1.1 are each defined to have dimensions that are not separated into subspaces. Thus there are no specified internal symmetry representations and no space-times.[8] There are no particles. There is no dynamics. There is no spatial separation—each space is a point space. Only their "essence" exists. See chapter 2.

1.3.2 Octonion Octonion Space

This space cannot be constructed by the annihilation of spaceless space fermions since no fermions exist in the spaceless spaces. See Fig. 1.2. This space has 4096 dimensions, which includes a 10 quaternion octonion space-time. It has urfermions with 32-spinors. It has internal symmetries and bosons that result from the upward extension of Megaverse space internal symmetries (that extend from QUeST universe internal symmetries). It has dynamical evolution. It can create Megaverse spaces through urfermion-antiurfermion annihilation. See chapter 3 for more details.

1.3.3 Quaternion Octonion Megaverse Spaces

Instances of this space can be constructed by the annihilation of octonion octonion space fermions. See Fig. 1.2. This space has 1024 dimensions, which includes an 8 complex octonion space-time. A Megaverse has a location and momentum in the octonion octonion space.-time. It has fermions with 16-spinors. It has internal symmetries that result from the upward extension of QUeST universe internal symmetries. It experiences dynamical evolution. It can create QUeST universe spaces through fermion-antifermion annihilation. See chapter 4 for more details.

1.3.4 Complex Octonion Universe Spaces

Instances of this space can be constructed by the annihilation of quaternion octonion space fermions. See Fig. 1.2. This space has 256 dimensions, which includes a 4 octonion space-time. A universe has a location and momentum in a Megaverse's space.-time. It contains fermions with 4-spinors. It has QUeST internal symmetries (consistent with UST) It experiences dynamical evolution. It can create minispaces through fermion-antifermion annihilation. See chapter 4.

1.3.5 Quaternion Point Universe Minispaces

Instances of this space can be constructed by the annihilation of complex octonion space fermions. See Fig. 1.2. This point space has no space-time, by construction, to preclude an infinite sequence of nested minispaces. A minispace has a location and momentum in a universe's space.-time. It has no fermions or bosons. It can have a subset of QUeST internal symmetries. It does not experience dynamical evolution. See chapter 5.

---

similar and may reflect a similarity of the Creator and physical Creation. (See *Sefer Yetzirah* – The Book of Creation.)

[8] We use the absence of space-times to end (truncate) the sequence of spaces in Fig. 1.1, that are connected by fermion-antifermion annihilation as discussed later.

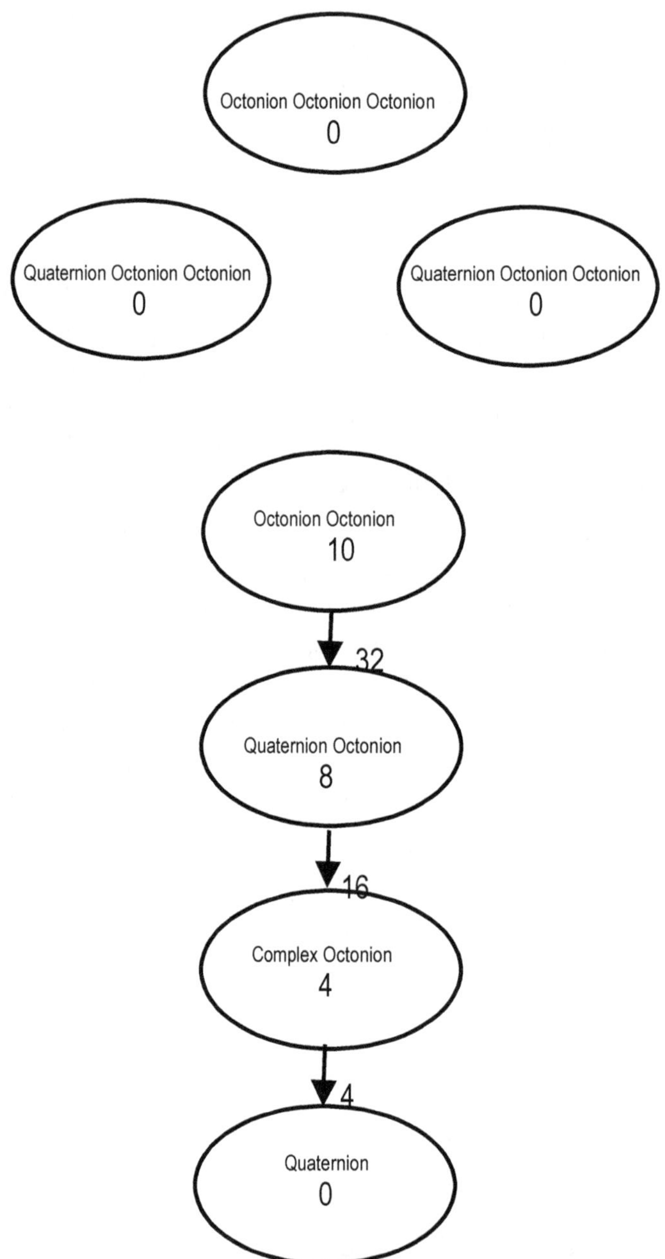

Figure 1.2. The seven spaces with their space-time dimensions indicated within each oval. The number of spinor components for each fermion - antifermion pair that annihilates to produce the "next space down" is specified next to each arrow for the four lower spaces.

1.4 Cosmological Evolution

The spectrum of spaces has an evolutionary character. The 3 spaceless spaces are static. They serve to close the set of spaces while maintaining the mathematical trend of coordinate octonions.

The five spaces: Complex Octonion Octonion, Octonion Octonion, Quaternion Octonion Megaverse, Complex Octonion Universe, and Quaternion Point Universe Minispace represent the spaces needed for the physical cosmos. The Complex Octonion Octonion and the Quaternion Point Universe Minispace mark the zero space-time dimension endpoints. These spaces are depicted in Fig. 1.3.

Turning now to the evolutionary sequence we find:

1. The Octonion Octonion Space must exist for all time. It is populated with elementary particles and interactions. After an unspecified time, an urfermion-antiurfermion annihilation may produce a Megaverse space. (There may be several/many Megaverses spread in time.)

2. A Megaverse space contains elementary particles and interactions. After an unspecified time a fermion-antifermion annihilation may produce a QUeST universe. There may be several/many universes over time.

3. A QUeST universe contains the fermions and interactions of the Unified SuperStandard Theory.

4. A fermion-antifermion annihilation can produce a minispace. Since a minispace has no space-time it can have no particles or interactions. It is empty. Yet it has topological significance. Minispaces have not been found experimentally. Considering their presumed rarity, and their lack of an experimental signature, the lack of experimental confirmation is understandable. If minispaces were detected they would be significant evidence for Octonion Cosmology!

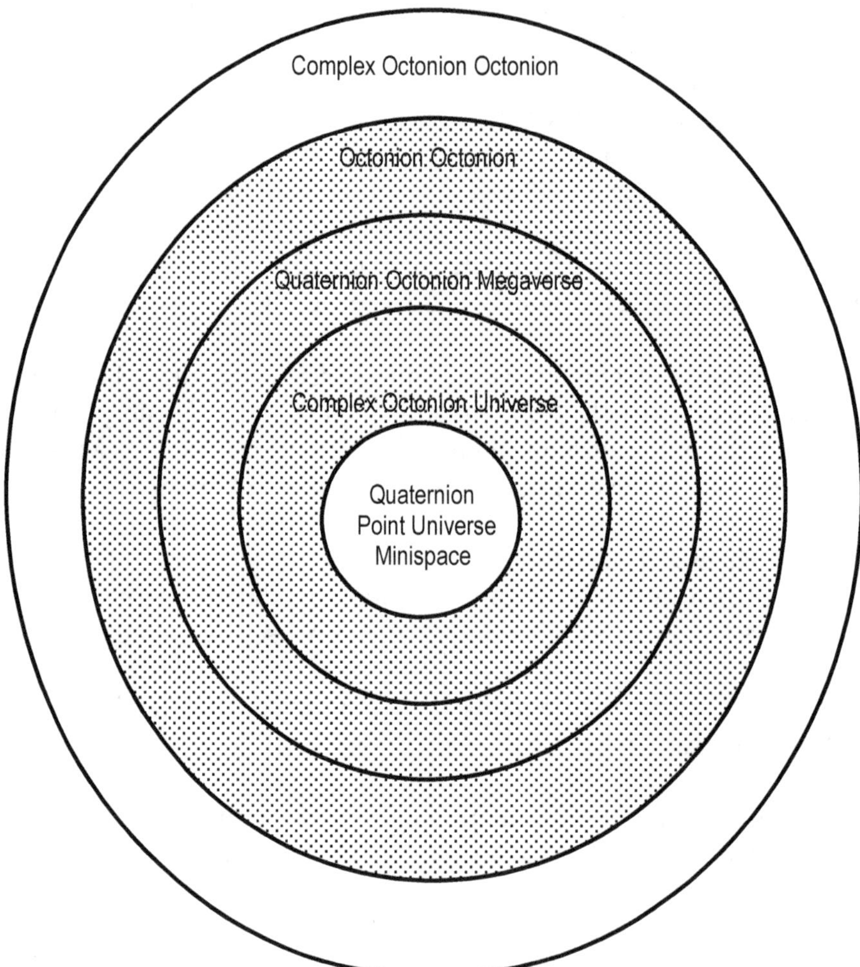

Figure 1.3. The five spaces relevant for the Cosmos. The three colored regions have non-zero dimension, and have a non-trivial space containing fermions and bosons with interactions.

2. The Three Spaceless Spaces

The three Spaceless Spaces are the Octonion Octonion Octonion, the Quaternion Octonion Octonion, and the Complex Octonion Octonion spaces. They have 512^2 = 262,144, 256^2 = 65,536 and 128^2 = 16,384 dimensions respectively. The spaces can be defined to have no space-time part. They could be defined to furnish fundamental representations for U(131072), U(32768), and U(8192) respectively. Since they have no dynamics, being timeless and spaceless, there can be no symmetry breaking. They have no elementary particles or interactions.

They can each be viewed as point spaces in the sense that they have no internal spatial distances.

They can also be viewed as exemplars for UST monads. We will consider monads in chapter 6.

3. Octonion Octonion Space: 10 Space-time Dimensions – With 32-Spinor urFermions

Octonion Octonion space is the ultimate source of Megaverses and universes in our Octonion Cosmology. This space has 64 rows of 64 dimensions. Its dimensions total to $64^2 = 4{,}096$.[9]

These dimensions are allocated to a ten quaternion octonion space-time (having 320 dimensions) with the remainder (4096 – 320 = 3776 dimensions) allocated to fundamental representations of internal symmetry groups.

The internal symmetry groups can be viewed as consisting of four copies of the internal symmetries of the UTMOST Megaverse internal symmetries PLUS[10] 512 – 320 = 192 dimensions allocated to additional internal symmetry groups. The 192 dimensions can be divided into four sets of 48 dimensions and added to each of the four copies. The 48 dimensions in each modified copy gives 12 dimensions to each layer of each copy.

Fig. 3.1 shows one layer of one copy of the UTMOST internal symmetry groups modified to include 12 additional dimensions. Octonion Octonion space has 16 copies of the internal symmetries in Fig. 3.1. The 12 dimensions can be initially allocated to a U(6), which may, in turn, be broken to the frequently found group U(1)⊗SU(2)⊗SU(3)⊗U(1) in QUeST[11] and UTMOST (or perhaps SU(3)⊗SU(3)).

The physical role of this space is to support the derivation of Megaverse spaces from urfermion-antiurfermion annihilation. The derivation is discussed in chapters 4 and 5 of *Cosmic seeds* ... later. Urfermions are 10 dimension fermions.

---

[9] The reader may benefit from reading the QUeST and UTMOST discussions of Appendices C and E of *Cosmic Seeds* ... , which appear later in this volume.

[10] The quantity $512 = 4 \cdot 128$ is the total number of space-time dimensions in four copies of UTMOST.

[11] See Figs. 3.2, 3.3, and 5.5 in *Cosmic Seeds* ... appearing later.

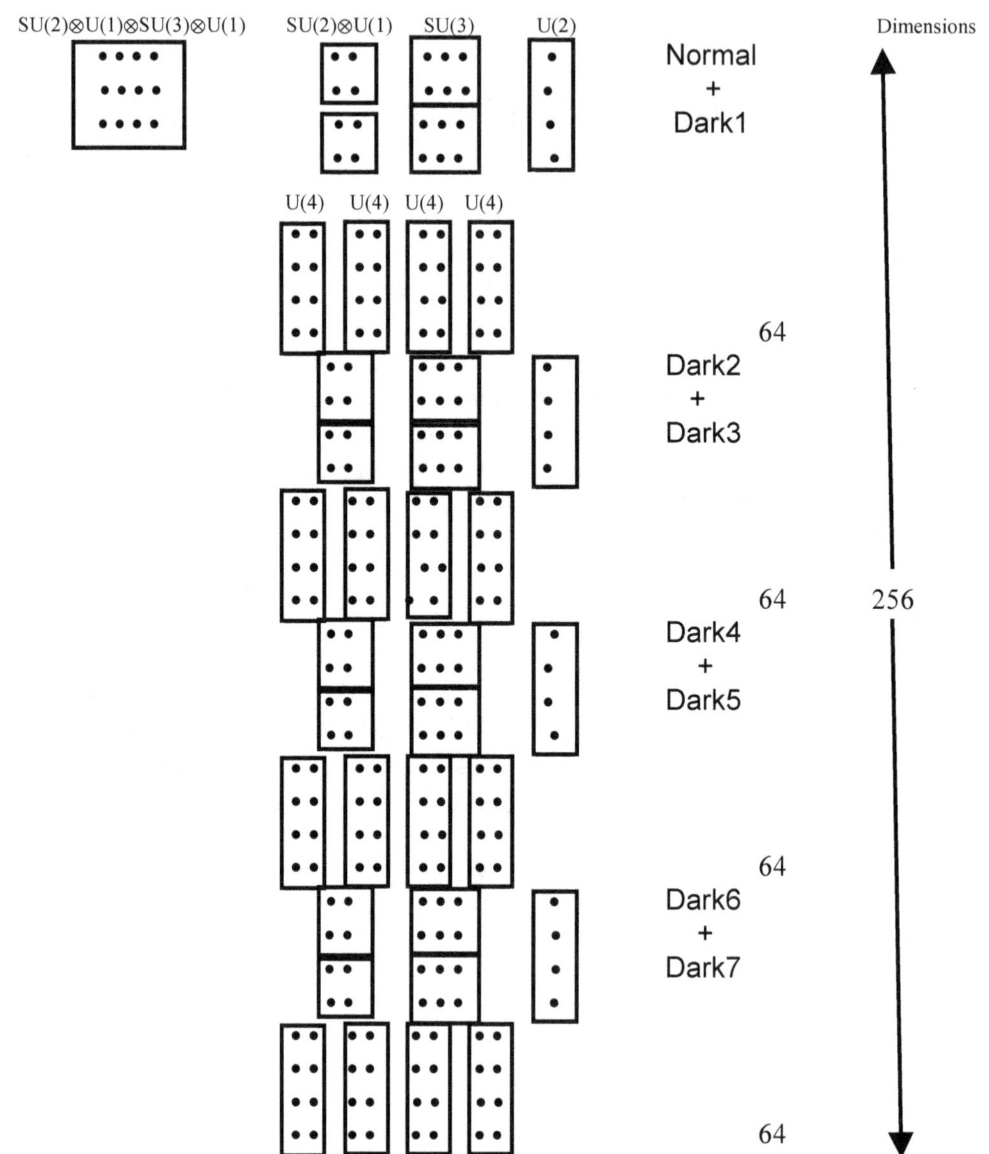

Figure 3.1 One layer of the four layers of one of the four copies of UTMOST dimensions *plus* an additional 12 dimension block with boxes around sets of dimensions for fundamental group representations. There are 16 copies of the internal symmetries listed here in Octonion Octonion space. Compare this figure to the UTMOST Fig. E.4 to see details of Normal and Dark sectors indicated above.

4. The UTMOST Megaverse and QUeST Universe Spaces

QUeST universe space is derived in chapters 1 and 2, and Appendix C of the book *Cosmic Seeds ...*, which is reprinted in the back of this book for the reader's convenience.

UTMOST Megaverse space is derived in chapters 4 and 5, and Appendix E of *Cosmic Seeds ....*[12] A reading of this book is recommended to better understand chapters 1 – 7.

[1212] We will reference items in *Cosmic Seeds ...* by this name.

5. The Quaternion Minispace

Instances of this space are generated by the annihilation of complex octonion QUeST fermion-antifermion pairs. QUeST fermions and antifermions have 4-spinors that are used to construct to construct the 4 × 4 dimension array for a Minispace. See Fig. 1.2. The construction is similar to those in chapters 1 and 2, and 4 and 5 of *Cosmic Seeds* .... We leave that as an exercise for the interested reader.

This point Minispace has no space-time by construction to preclude an infinite sequence of further nested minispaces. A minispace has a location and momentum in a universe's space.-time. It has no fermions or bosons. It has a subset of QUeST internal symmetries (perhaps U(1)⊗SU(2)⊗SU(3)⊗U(1)), which are not instantiated with vector bosons. It does not experience dynamical evolution.

An instance of this space could be detected experimentally as an annihilation event with no final state particle produced that ultimately decays into a fermion-antifermion pair. Or an experiment may see the creation of a fermion-antifermion pair from "nothing." No evidence for these phenomena has been found to the author's knowledge.

6. Monads of Elementary Particles – Extension to 10 Spaces

In previous books[13] the author has shown that the relativistic problem of instantaneous communication between parts of a quantum system at space-like distances can be resolved through the factoring of quantum wave functions into a functional and a spatial wave. Functionals, which embody the instantaneous effects, were in a point space of functionals. Spatial waves were in a point Fourier wave space. Thus instantaneous effects did not proceed through coordinate space distances. See Appendix 6-A.

The general form of a Dirac fermion wave function is

$$\psi(x) = \sum_{\pm s} \int d^3p N(p)[b(p, s)u(p, s)e^{-ip\cdot x} + d^\dagger(p, s)v(p, s)e^{+ip\cdot x}] \tag{6.1}$$

It can be “factored” into a functional inner product formalism in the manner of Riesz (1955)[14] and others:

$$\psi(x) = (f, (s, x, t)) \tag{6.2}$$

In the general case a functional for a single particle state can be expressed as a “product” of factors:

$$f_{IS,spin,momentum} = f_{IS}\, f_{spin,\ momentum}$$

where f_{IS} is a functional specifying the Internal Symmetry, and $f_{spin,\ momentum}$ is a functional specifying the spin and momentum.

We identify the functionals defined above as *monads*. In Blaha (2020c) and earlier books we showed that we can take every elementary particle to have a monad within it. Fermion monads, called *qubes*, appear to give a small mass to fermions. We suggest this mass was of the order of the *Landauer energy*. Boson particle monad (*Quba*) masses were taken to be zero in order to avoid conflicts with gauge symmetries.

Justification for the “factoring” of wave functions into functionals and waves is presented in Appendix 6-B.

6.1 Internal Symmetry Functionals Space F_{IS}

We can view the internal symmetry functionals as resident in a space of functionals. This space has zero distance similar to the spaceless spaces considered in chapter 1. Such a space has a functional for each fermion and boson in QUeST and UST. This space can be extended to handle multi-particle states by supporting products

[13] See for example Blaha (2020c) and references therein.

[14] For example see pp. 61-2 of Riesz (1955) where linear functionals and their inner products are defined.

and sums of individual particle functionals. It is infinite since it must support all possible particle configurations.

There is no distance in the set of functionals. Thus, as we showed in Blaha (2020c) and earlier books, factoring quantum fields into inner products of functionals and waves eliminates the instantaneity problem of Quantum Entanglement.

We will denote this point space as F_{IS}.

6.2 Spin-Momentum Functionals Space F_{SP}

We can define another infinite space consisting of functionals for all possible spin and momentum states. This space has no distance measure and can be viewed as a point space. We will denote this point space as F_{SP}.

6.3 Fourier Wave Space F_{FW}

We can define a space of all possible Fourier waves F_{FW}. This space also has no distance measure and can be viewed as a point space.

6.4 Space-Time and Functional Spaces

At each point in a universe's space-time, we can view the three point spaces above as present. They exist as a "tensor product" with the space-time. There is only one of each point space, which is shared by each space-time point.

A spatially dispersed quantum system has instantaneous access to its composite functional. *Thus a change in one part of the system causes instantaneous changes in other parts.*

6.5 Ten Spaces of Octonion Cosmology

The seven octonion spaces introduced in chapter 1, plus the three functional spaces presented here, gives us a Ten Space Cosmology.

This chapter was presented within the framework of free fields and their products. It is general because all measurements are made on in-states and out-states. Quantum measurements within interaction regions are generally not possible.

6.6 Dynamic Generation of Quantum State Functionals

The preparation of a physical system of particles is always required to be local. After preparation, a quantum system can evolve to have widely separated parts at space-like distances.

The preparation of a quantum system requires functionals for each of the parts (particles) within it. These functionals are generated by dynamic extensions (and combinations) of the functionals of the three spaces defined above. Thus the above three spaces must be extendable—without limit—to accommodate all possible physical state preparations.

In the preparation of a state, the functionals of different parts of the state are interrelated leading to effects such as quantum entanglement.

Appendix 6-A. Quantum Entanglement and Action-at-a-Distance

Einstein, Podolsky, and Rosen (EPR) considered the quantum entanglement of two systems and showed that instantaneous action-at-a-distance (spookiness) resulted. In this appendix we will show that our quantum functional formalism, which generalizes quantum theory, eliminates the problem of instantaneous action-at-a-distance.[15] We will show the solution provided by quantum functionals using the same example as EPR.

The key feature of quantum functionals is their ubiquitous presence at every point of space-time. In a multi-system quantum state all functionals are directly, instantaneously linked no matter what the separation of the constituent systems. When a reduction of part of the state of a system occurs due to a measurement, all other parts are instantly updated since the space-time separation of the individual parts is not relevant. The linkage of all quantum functionals is relevant. A reduction of one part 8.immediately impacts the other related parts.

6-A.1 The EPR Two System State Example

EPR considered a state consisting of two systems that might become separated spatially. We can represent the state as

$$\Psi = \Sigma_n \psi_{1n}(x_1)\psi_{2n}(x_2) \qquad (6\text{-A.1})$$

We can represent a measurement (reduction of state) with a projection Π_{1a} of system "1" to a state ψ_{1a} with

$$\psi_{1a} = \delta_{ab}\, \Pi_a\, \psi_{1b} \qquad (6\text{-A.2})$$

Then

$$\Psi_{\text{projected}} = \Pi_{1a}\, \Sigma_n \psi_{1n}\psi_{2n} = \psi_{1a}(x_1)\psi_{2a}(x_2) \qquad (6\text{-A.3})$$

The effect of the measurement of system "1" is *instantaneous* on system "2" because the quantum functionals f_{1n} and f_{2n}, and the projections Π_{1n} and Π_{2n} of both systems are not separated by distance. Thus

$$\psi_{1n}(x) = f_{1xn}(\Pi_{1xn}\Phi) = (f_{1xn}, \Pi_{1xn}\Phi) \qquad (6\text{-A.4})$$

$$\psi_{2n}(y) = f_{2ny}(\Pi_{2yn}\Phi) = (f_{2yn}, \Pi_{2yn}\Phi) \qquad (6\text{-A5})$$

[15] This section presents the solution for quantum spookiness that we proposed in Blaha (2019g) and (2018e).

with $x = x_1$ and $y = x_2$. *The quantum functional and the projection select the wave and its coordinate parameterization. The coordinates in the wave are merely place holders.*

Therefore the relative distance between the coordinates x_1 and x_2 is not relevant for the change of state of system "2". The quantum functionals and projections give the instantaneity of the change in ψ_{2a} upon the measurement of system "1".

EPR Spookiness is resolved by quantum functionals. There is no conflict with the Theory of Special Relativity.

Appendix 6-B. Why Factorize Quantum Fields into a Functional and a Fourier Expansion?

We identify three reasons for factoring quantum fields:

1. The internal symmetries of free fundamental fermions are only partially correlated with their fourier expansions: there is a spin/handedness relation and a species relationship based on the species of each fundamental fermion. The internal symmetries: color, generation number, layer number, and so on appear only as indices attached to creation and annihilation operators. Functionals are matched with corresponding fourier wave expansions using functional inner products.

2. Quantum Entanglement between parts of a physical state separated by space-like distances suggest the correlated changes in the quantum numbers of a physical state with spatially separated parts are independent of spatial distance. Quantum number changes (in any quantum numbers) between spatially separated parts are instantaneous. This fact demonstrates phenomena strongly supports the factorization of the quantum numbers and the coordinates of quantum fields.

3. The close analogy between the subgroups of the Complex Lorentz group and the known symmetries of The Standard Model suggest that the internal symmetries of fermions represented by functionals have a somewhat similar group structure to space-time symmetries.

For these reasons we propose to consider quantum fields as inner products of functionals and space-time fourier waves.

7. Basis of Unified SuperStandard Theory in Octonion Spaces

Chapter 3 of *Cosmic Seeds* ... following describes the connection of the QUeST octonion space to the Unified SuperStandard Theory (UST). The space-time, limited to real coordinates, is the same; the internal symmetry groups are the same; and features such as monads can be derived from the 10 space formulation of Chapter 6.

8. Octonion Cosmology

This volume presents the panorama of Octonion and Functional Spaces needed for a Physical Cosmology at the deepest level of Physical Reality. It is difficult to envision a deeper physical level.

It justifies the Unified SuperStandard Theory (UST) that the author developed in the past twenty years based on the method of Euclid's Logic. UST contains the Standard Model of Elementary Particles within the framework of an extended Quantum Field Theory that originates in the author's papers of the 1970's.

The beauty of the octonion spectrum of spaces presented in Fig. 1.1 argues for its elegance and simplicity as well as its fundamental significance. The directness of the derivation of UST, and its close connection to experimental reality as far as we know it, suggests it has the general framework needed to understand elementary particle physics in detail.

The Cosmic Seeds of Universes and Megaverses

The Cosmic Seeds of Universes & Megaverses

Stephen Blaha Ph. D.
Blaha Research

Towards The Deepest Level of Physical Reality
Detailed Derivation of a QUeST Universe from a One *Megaverse* Fermion
Detailed Derivation of the UTMOST Megaverse from a 10 Dimension Fermion
Evidence The Seed Fermion Spinors Structure The Universe Dimension Array—QUeST
Evidence that Megaverse Seed Fermion Spinors Structure Its Dimension Array
Derivation of UST from QUeST
Interconnection of Our Universe and the Megaverse

Pingree-Hill Publishing
MMXX

ISBN: 978-1-7356795-1-8

Rev. 00/00/01 October 14, 2020

INTRODUCTION

The past 100 years have witnessed major growth in our understanding of elementary particles and Physical Reality. At this point in time it is appropriate to create a synthesis of the cumulative acquired physical knowledge. This book combines the work of the author over the past fifty years to develop a synthesis. The result is a complete theory of Elementary Particles and Gravitation centered on a generalization of space that combines internal symmetries and space-time in a simple theory built ultimately on a 10 dimension space. The space supports a primitive type of fermion called an urfermion. This fermion and its antiparticle annihilate to produce the 1024 dimension Megaverse. The Megaverse in turn supports seed fermions. A seed fermion and its antiparticle can annihilate to generate a universe particle containing a QUeST (Quaternion Unified SuperStandard Theory) universe with 256 dimensions.

A QUeST universe contains the internal symmetries of the UST (Unified SuperStandard Theory) and a four complex quaternion dimension space-time. UST contains the Standard Model as well as additional reasonable symmetries.

The result is a complete theory! Blaha (2020c) and its predecessors describe UST in detail. A series of books this year by the author describes the deeper basis of UST in QUeST, BQuEST, UTMOST, and BMOST.

The Quaternion Unified SuperStandard Theory (QUeST) is a theory for universes that is based on a thirty-two complex quaternion space. It implies the Unified SuperStandard Theory (UST) of the author to the author's initial surprise. The Megaverse theory UTMOST (Megaverse Octonion SuperStandard Theory) is based on a sixty-four complex octonion space. Both theories are based on deeper theories BQUeST and BMOST described by this book.

This book provides a detailed derivation of QUeST from a Megaverse fermion theory and describes the origin of universes as particles. It also provides a detailed derivation of UTMOST from a 10 dimension fermion theory and provides details of the origin of the Megaverse as a particle. It reduces the origin of universes and of the Megaverse to the simplest possible basis. An important aspect of these derivations is the interrelation of universes with the Megaverse. The one fermion seed for universes is in the Megaverse; the one fermion seed for the Megaverse is in a ten dimension space that may be a SuperSymmetry space.

This book also proposes a clear multilevel sequential patterns of symmetry breaking of QUeST down to the level of SU(3) and SU(2)⊗U(1) symmetries. It also considers the detailed derivation of UST from QUeST, and provides an .Overview of QUeST and its UST sector features. An Overview of Megaverse UTMOST features is also presented.

This book completes the panorama of hypercomplex features for QUeST universes and the UTMOST Megaverse. The result is a complete theoretical framework for Elementary Particles and Gravitation.

In the interests of completeness the book contains five appendices detailing aspects of the Unified SuperStandard Theory (UST), QUeST, and UTMOST for use in derivations.

1. The Derivation of Quaternion QUeST Universes

QUeST (and the Unified SuperStandard Theory that it implies) is based on a 32 dimension complex quaternion space (256 total dimensions). The questions arise: Is there a deeper basis for QUeST with much fewer dimensions? Can QUeST, which describes a universe, be a subspace that originates as a "particle" in a larger space that we call the Megaverse (Multiverse)? Questions of this sort have led the author to find a deeper theory that implies 256 dimension QUeST. The theory, called BQUeST, is described in Blaha (2020i) and (2020j).

This chapter describes BQUeST in much greater detail. The process for the birth of a universe begins in one fermion located in the 8-dimension Megaverse. *The fermion, which is off shell, has 16 independent spinor components that lead to the 256 internal dimensions of a QUeST universe.* Appendix C describes features of the QUeST quaternion space.

1.1 BQUeST Origin

We begin with the assumption that a QUeST universe originates in the UTMOST Megaverse.[16] Since the UTMOST Megaverse has 8 space-time dimensions, an eight dimension seed fermion has 16 spinor components. (There are 256 = 16×16 Dirac matrices, denoted G_{ab} for a, b = 1, 16, and eight 16 × 16 Dirac γ matrices.)

We view the Megaverse as the location of the origin of a QUeST universe. We picture an off shell seed fermion and its off shell seed antifermion as appearing as fluctuations in the Megaverse vacuum. They "annihilate" to create a QUeST universe particle.[17] The universe particle, being in a Big Bang state of enormous energy, then proceeds to expand.

We will show the 16 independent spinor components of the 8 dimension off shell seed fermion and of the off shell anti-seed fermion generate the 16 × 16 dimension array of QUeST.[18]

1.2 A Model Dynamics of the Seed – Anti-Seed Generation of a QUeST Universe

The seed fermion must have a dynamics that enables it to generate the 256 dimension array. In this section we describe a model dynamics for the annihilation of a seed–antiseed state into a scalar universe particle. The model must have certain features that accomplish the goal:

---

[16] QUeST universes are described in Appendix C. The Utmost Megaverse is described in Appendix E.

[17] Appendix D provides experimental and theoretical evidence for a particle view of universes. Comments by DeWitt and others on quantum universes support this view.

[18] The UTMOST Megaverse has an eight complex octonion space-time. We choose to extract the real coordinate of each complex octonion 16-tuple to construct a real 8 dimension space-time within which seed fermion dynamics occurs. This is analogous to the case of complex 4 dimension space-time being restricted to real complex space-time in conventional quantum field theory. Streater (2000) maintains complex space-time is needed for axiomatic quantum field theory.

1. Since the target dimension array is not symmetric the PseudoQuantum formulation[19] of Quantum Field Theory will be seen to be required.

2. Since the quantum field theory is in 8 dimensions we must use Two-Tier coordinates,[20] which eliminate all divergences in perturbation theory using exponential damping of all integrations.

These extensions of quantum field theory were developed by the author in the 1970's and the early 2000's. They are both described in Appendix A together with references to earlier papers and work.

We define a model lagrangian for the seed fermion sector. The seed fermion is represented by *two* quantum fields[21] ψ_1 and ψ_2. The scalar[22] universe particle has two quantum fields A_{1ab} and A_{2ab} with indices a and b.

The lagrangian terms are

$$\mathscr{L} = F^{1\mu\nu} F^{2}{}_{\mu\nu} + \overline{\psi}_2 (i\gamma\cdot\partial - m)\psi_1 + \overline{\psi}_1 (i\gamma\cdot\partial - m)\psi_2 - e_0\overline{\psi}_2 A_2\psi_1 - \overline{\psi}_1 A_1\psi_2 \tag{1.1}$$

where m is the mass of the seed fermion, and where $A_i = A_{iab}G_{ab}$ summed over a and b.

To insure finite perturbation series results (in 8 dimensions) the fields are all functions of Two-Tier coordinates X having a vector quantum field $Y^\mu(y)$:

$$X^\mu(y) = y^\mu + i\, Y^\mu(y)/M_c^{\,2} \tag{1.2}$$

where M_c is an extremely large mass.

1.2.1 U(16) Species Group Symmetry for Dirac Matrix Indices

The lagrangian terms (plus additional universe terms) are invariant under a U(16) Species symmetry group. Appendix B describes the analogous U(4) Species[23] symmetry for four dimension fermions that we found for UST.

Appendix B shows the form of the trial lagrangian for establishing the Species group of UST in conventional quantum field theory:

$$\overline{\psi}'(x)[i\gamma_\mu'(x)(\partial/\partial x_\mu - ig_8A'_{Sk}{}^{\mu}(x)G_{Sk}) - m]\psi'(x) \tag{51.2}$$

---

[19] S. Blaha, Il Nuovo Cimento **49A**, 35 (1979) and Appendix A.

[20] See Appendix A.

[21] We use a PseudoQuantum scalar pair of fields also. A scalar particle is a model of the universe. See eqs. 61 – 90 in Appendix 1-C of Blaha (2020j). The Two-Tier formulation eliminates all perturbation theory divergences in four dimensions including the notorious fermion triangle divergence. It also eliminates all perturbation theory divergences in eight dimensions.

[22] Current experimental evidence suggests the universe is not rotating.

[23] The U(4) Species symmetry group is part of UST. It originates in the Complex General Reloativity of four dimension coordinates as shown in Appendix A and Blaha (2020c) as well as earlier books.

where the matrices G_{Sk} are the 16 Dirac matrices of 4 dimension space, where k ranges from 1 through 16, and where the $A'_{Sk}{}^{\mu}$ fields.are 16 gauge fields for the Species group.

In developing the U(4) Species group symmetry for four dimension UST we related the indices of the Dirac matrices $[G_{Sk}]_{ab}$ where a, b = 1, 2, 3, 4 to the space-time coordinates using vierbein transformations. In this regard Appendix B states:

A complex transformation of types II and III in Fig. 50.1 has the form:

$$U(x'')^{\mu}{}_{\nu} = w^{\mu}{}_{a}(x'')[\exp(i\sum_k \Phi_k(x'')\tau_k)]^{a}{}_{b}\, l^{b}{}_{\nu}(x'')$$
$$U^{-1}(x'')^{\mu}{}_{\nu} = w^{\mu}{}_{a}(x'')[\exp(-i\sum_k \Phi_k(x'')\tau_k)]^{a}{}_{b}\, l^{b}{}_{\nu}(x'')$$

where τ_k is a U(4) generator matrix. Its infinitesimal transformation is approximately

$$U(x'')^{\nu}{}_{\beta} \approx \delta^{\nu}{}_{\beta} + i\sum_k \Phi_k(x'')[\tau_k]^{\nu}{}_{\beta} \qquad (50.12)$$
$$U^{-1}(x'')^{\nu}{}_{\beta} \approx \delta^{\nu}{}_{\beta} - i\sum_k \Phi_k(x'')[\tau_k]^{\nu}{}_{\beta}$$

using the *vierbein* flat space-time limits

$$w^{\mu}{}_{a}(x'') \approx \delta^{\mu}{}_{a} \qquad (50.12a)$$
$$l^{b}{}_{\nu}(x'') \approx \delta^{b}{}_{\nu}$$

where

$$\Phi_k(x) = \int^{x} dy_{\lambda}\, A_{Rk}{}^{\lambda}(y) \qquad (50.13)$$

Then

$$\Gamma_{R}{}^{\sigma}{}_{\lambda\mu} = -\tfrac{1}{2}i\{\sum_k A_{Rk}(x'')_{\mu}[\tau_k]^{\sigma}{}_{\lambda} + \sum_k A_{Rk}(x'')_{\lambda}[\tau_k]^{\sigma}{}_{\mu}\} \qquad (50.14)$$
$$- A_{R}{}^{\sigma}{}_{\mu\lambda} + A_{R}{}^{\sigma}{}_{\lambda\mu}$$

(summed over k) with the matrix $A_{R}{}^{\sigma}{}_{\mu\lambda}$ given by

$$A_{R}{}^{\sigma}{}_{\mu\lambda} = -\tfrac{1}{2}i\sum_k A_{Rk\mu}[\tau_k]^{\sigma}{}_{\lambda} \qquad (50.15)$$

with $A_{R}{}^{\sigma}{}_{\mu\lambda}$ transformable to matrix row and column numbers

$$A_{Rflat}{}^{\mu a}{}_{b} = A_{Rflatk}{}^{\mu}[\tau_k]^{\sigma}{}_{\lambda}\delta_{\sigma}{}^{a}\delta^{\lambda}{}_{b} \qquad (1.3)$$

using the flat space-time vierbein values, and so $A_{Rflat}{}^{a}{}_{\mu b}$ may be written in matrix form as

$$A_{Rflat\mu} = -\tfrac{1}{2}i\sum_k A_{Rflatk\mu}\tau_k \qquad (50.16)$$

In the flat spacc-time limit the $A_{Rk}{}^{\lambda}(y)$ becomes the Coordinate Species group U(4) gauge fields $A_{Rflatk}{}^{\lambda}(y)$.

The above extract shows that space-time dimensions can be directly related to matrix indices. We will use this feature to map the indices of the universe fields A_{iab} to internal dimensions of the universe particles. These dimensions, which number 256 in the 8 dimension Megaverse space-time, become the 256 dimensions in QUeST. (It should be remembered that the UTMOST Megaverse has 1024 dimensions —much more than the 256 dimensions of QUeST.

1.3 The Interaction Generating a Universe Particle

The diagram for the creation process is in Fig. 1.1. A seed fermion and an antiseed fermion annihilate to create a scalar universe particle.[24]

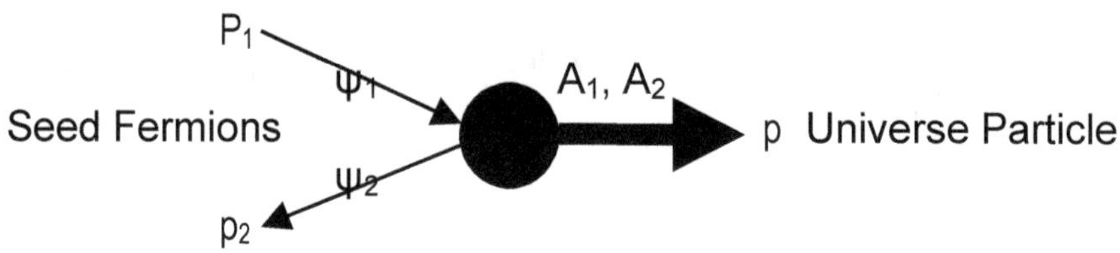

Figure 1.1. Diagram for the transition of seed - antiseed fermions to a universe particle.

The lowest order S-matrix element for this interaction has the form:

$$S = \int d^8x\, [\overline{\psi}_2 A_2 \psi_1 + \overline{\psi}_1 A_1 \psi_2] \tag{1.4}$$

After evaluating the integration we find S has terms of the form

$$\overline{v}_{2a}(p_2)\, A_{1ab}(p)\, u_{1b}(p_1) \tag{1.5}$$

Rearranging the expression we obtain the trace expression

$$\text{Tr}\ u_{1b}(p_1)\overline{v}_{2a}\,(p_2)\, A_{1ab}(p) = U_{ba} A_{1ab}(p) \tag{1.6}$$

where U_{ba} is an array composed of u and $\overline{v}$.

Note the interaction can only take place for far off shell seed and antiseed fermions. Then the 16 spinor components of each fermion are independent and can be treated as independent.

Universe production (Fig. 1.1) has momentum conservation due to the off shell fermions. The diagram has two interesting cases:

---

[24] The possibility of seed-antiseed annihilation into two universe particles is not excluded This possibility might account for the handedness and preponderance of matter (as opposed to antimatter) in our universe.

1. If A_{1ab} and A_{2ab} are both proportional to a Kronecker δ_{ab} then the S matrix expression of eq. 1.4 simplifies to the scalar A_i case.
2. If A_{1ab} and A_{2ab} are not proportional to a Kronecker δ_{ab} then the S matrix expression can be expressed as a sum of terms for each value of a and b:

$$S = \Sigma\, S_{ab} \tag{1.7}$$

where (with no sum over a or b)

$$S_{ab} = \int d^8x\, [\bar{\psi}_{2a}A_{2ab}\psi_{1b} + \bar{\psi}_{1a}A_{1ab}\psi_{2b}] \tag{1.8}$$

The second expression eq. 1.8 suggests each ab pair of values marks a channel from the fermions to the universe particle. Thus we can view the universe particle as receiving 256 “streams” from the fermion-antifermion pair.

In section 1.4 below we suggest the streams “amalgamate” to form a 256 dimension universe.

1.4 U(16) Symmetry of A_i

The lagrangian in eq. 1.1 has a U(16) symmetry that is analogous to the 4 dimension U(4) Species group described in section 51.1 of Appendix B. This U(16) symmetry exists in spinor “space.” The matrix, $[U_{ba}]$ with its 16-spinor indices a and b, has a symmetry under the U(16) transformation:

$$U[U_{ba}]U^{-1} \tag{1.9}$$

U(16) symmetry requires the A_i factors in eqs. 1.5 and 1.6 to also have a U(16) transformation of the form:

$$A_i' = UA_iU^{-1} \tag{1.9a}$$

This symmetry will be seen to impact on the form of the QUeST dimension array.in chapter 2.

1.5 Indices of $[A_{1ab}]$ Transformed to Internal Dimensions

Eq. 1.5 shows the indices of the array, $[A_{1ab}]$, have $16 \times 16 = 256$ values due to the 16×16 independent values of the products: $u_{1b}(p_1)\bar{v}_{2a}(p_2)$. Eq. 1.3 above shows that indices can be transformed to coordinates.

Therefore we can take the array of indices [ab] and transform them to coordinates using a Megaverse vierbein.[25] We can take the coordinates that emerge as each corresponding to a one dimension line (a stream?) within the universe particle. Subsequently, or simultaneously, after creation the energy-momentum within the streams it can be distributed/flow into the 4 dimension space-time within the QUeST universe. The internal energy-momentum must sum to zero for momentum

[25] The Megaverse undoubtedly has an 8-dimension General Relativity. For small neighborhoods of a point the vierbein becomes the identity as eq. 50.12a above shows.

conservation. So there will be patches of positive and negative energies. (Negative energies may be absorbed by a vacuum rearrangement.)

The independence of the spinor components of each of the two seed fermion fields guarantees the elements of the array of indices form an array of independent dimensions.

1.6 The Difference between Indices and Internal Dimensions

The array $[A_{1ab}]$ is an array with indices. We now consider the relation of indices and internal dimensions of an entity. First we consider the case of a cube, which can be viewed as an entity with three indices. Its external indices can be taken to label the length, width and height of the cube. Within the cube we can establish a coordinate system with three dimensions labeling points within it. Thus external indices can be viewed as mapping to internal dimensions within.

Perhaps a better example is a Black Hole. It exists in 4-dimension space. Within its event horizon the coordinates of the external space can be continued. It is also possible to replace them with an internal coordinate system of four dimensions. This possibility could be supported by the role of radial dimension externally becoming effectively a time coordinate internally.

We thus conclude the indices of A_{iab} can be viewed as specifying coordinates (*dimensions)* within A_i. We have mapped the seed fermion to a $16 \times 16 = 256$ dimension array. This array serves as the dimensions of the QUeST space.

1.7 Origin of Universe

We can then envision the possibility that our universe started as the annihilation of a seed fermion and antifermion in the Megaverse. It "acquired" the 256 dimensions of QUeST to form a space that includes our 3+1 space-time at the Big Bang point as well as internal symmetries. This process could proceed, as it likely does, instantaneously. The seed fermion-antifermion pair are necessarily in an 8-dimension UTMOST space-time as we showed in previous books Thus QUeST universes are directly linked to the UTMOST Megaverse.

Appendix D shows that the post-Big Bang stage of the universe is analogous to the vacuum polarization of a charged particle, substantiating the interpretation of the universe as a type of particle.

1.8 Nature of the Seed Fermion and Antifermion

The type of the seed fermion and antifermion are one of the possible 1024 fermions in the UTMOST Megaverse. In Blaha (2020e) we showed that QUeST and MOST fermions occur in four species: charged leptons, neutral leptons, up-type quarks, and down-type quarks. Examples are e, ν_e, u, and d. The same set of four species appears in UTMOST as shown in section 1.9 below.

There are four layers of fermions in UTMOST. Each layers has eight sets of four generations of four species. In total there are $128 \cdot 8 = 1024$ fermions. A reasonable possibility is the seed fermion is an *electron-type fermion* with the lowest mass of all 1024 fermions based on the masses of the known fermions.

1.9 Demonstration That UTMOST Fermions Occur in Four Species

This section follows Blaha (2020e) applied to UTMOST.

In the Megaverse, the fermion species in UTMOST number four despite the 8-dimension complex octonion nature of UTMOST space-time.

There are four types of boosts in this 7+1 complex octonion valued space-time that boost a particle rest state to a state of motion with a real energy, and a real-valued or complex octonion valued momentum (spatial coordinates):

1. A boost from rest to a frame with real-valued energy and momentum with $p^{02} - ||\mathbf{p}||^2 > 0$. A "normal charged lepton-like" fermion.

2. A boost from rest to a frame with real-valued energy and momentum with $p^{02} - ||\mathbf{p}||^2 < 0$. A "tachyonic neutral lepton-like" fermion.

3. A boost from rest to a frame with real-valued energy and a complex quaternion valued spatial 7-momentum with $p^{02} - ||\mathbf{p}||^2 > 0$. An "up-type quark-like" fermion.

4. A boost from rest to a frame with real-valued energy and a complex quaternion valued spatial 7-momentum with $p^{02} - ||\mathbf{p}||^2 < 0$. A "tachyonic down-type quark-like" fermion.

where p^0 is the real part of the octonion energy, and where $\mathbf{p}$ is a 7-vector and each of its momentum components $\mathbf{p_k}$ are spatial 16-vector (complex octonion):

$$\mathbf{p} = \mathbf{p}_1 + \mathbf{p}_2 + \mathbf{p}_3 + \ldots + \mathbf{p}_7$$
$$\mathbf{p}_k = \mathbf{p}_{kr} + i\mathbf{p}_{ki} + j\mathbf{p}_{k3} + k\mathbf{p}_{k4} + q\mathbf{p}_{k5} + r\mathbf{p}_{k6} + s\mathbf{p}_{k7} + \ldots + t\mathbf{p}_{k16} \quad (1.10)$$
$$||\mathbf{p}|| = \text{sqrt}(\mathbf{p}_r\cdot\mathbf{p}_r + \mathbf{p}_i\cdot\mathbf{p}_i + \mathbf{p}_3\cdot\mathbf{p}_3 + \ldots + \mathbf{p}_{16}\cdot\mathbf{p}_{16})$$

where i, j, k, q, r, s, …, and t are fundamental octonion units, and where $||\mathbf{p}||$ is the norm of $\mathbf{p}$. The $\mathbf{p}_k\cdot\mathbf{p}_k$ terms are 16-vector inner products.

2. Connection of Seed Fermion Structure to QUeST Dimensions Structure

Decisive experimental evidence for the origin of a universe in seed fermion-antifermion annihilation will not be found for many millenniums, if at all. There is a possibility that theoretical support for our proposed origin of universes may arise. The possibility is based on the possibility that the structure of the QUeST dimension array have the "imprint" of the form of the seed fermion-antifermion prior to annihilation.

The hope: The seed fermion-antifermion annihilation that produces a QUeST universe has features that appear to impress themselves on universe structure. This chapter describes this possibility.

2.1 Structure of 16-Spinor u and v

The sixteen 16-spinors of an 8 dimension fermion have the general form:

Number of Columns:	4	4	4	4
	u-type fermion)		v-type (anti-fermion)	
4	u spin up		v small terms 1	
4		u spin down	v small terms 2	
4	u small terms 1		v spin down	
4	u small terms 2			v spin up

Figure 2.1. The 16 spinors of a 8 dimension spinor. Each spinor has 16 rows.

2.2 Compression of u and v into Composite Spinors

We can combine the eight u-type spinors into a composite spinor to exhibit their structure. Similarly, we can combine the eight v-type spinors into a composite spinor. Fig. 2.2 shows the forms of the composite u-type and v-type spinors.

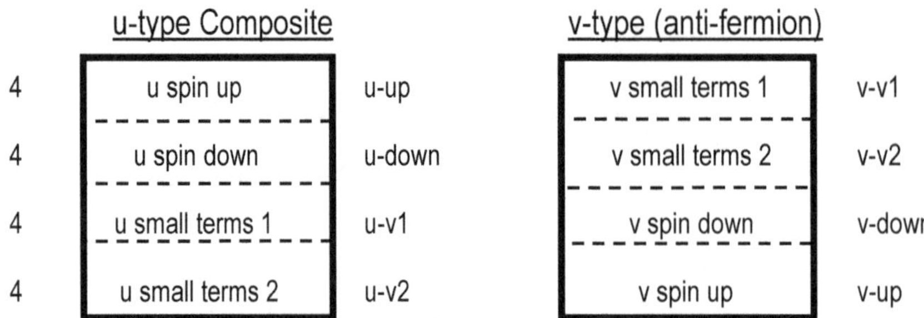

Figure 2.2. Composite u-type and v-type spinors illustrating the structure of each type of spinor. Each spinor is subdivided into 4 row sections. The third and fifth columns are symbols for each sector.

The outer product of spinor factors in eq. 1.6 is u$\bar{v}$. Fig. 2.3 contains the form implied by the composite spinors.

u-up–v-v1	u-up–v-v2	u-up–v-down	u-up–v-up
u-up–v-v1	u-up–v-v2	u-up–v-down	u-up–v-up
u-up–v-v1	u-up–v-v2	u-up–v-down	u-up–v-up
u-up–v-v1	u-up–v-v2	u-up–v-down	u-up–v-up
u-down–v-v1	u-down–v-v2	u-down–v-down	u-down–v-up
u-down–v-v1	u-down–v-v2	u-down–v-down	u-down–v-up
u-down–v-v1	u-down–v-v2	u-down–v-down	u-down–v-up
u-down–v-v1	u-down–v-v2	u-down–v-down	u-down–v-up
u-v1–v-v1	u-v1–v-v2	u-v1–v-down	u-v1–v-up
u-v1–v-v1	u-v1–v-v2	u-v1–v-down	u-v1–v-up
u-v1–v-v1	u-v1–v-v2	u-v1–v-down	u-v1–v-up
u-v1–v-v1	u-v1–v-v2	u-v1–v-down	u-v1–v-up
u-v2–v-v1	u-v2–v-v2	u-v2–v-down	u-v2–v-up
u-v2–v-v1	u-v2–v-v2	u-v2–v-down	u-v2–v-up
u-v2–v-v1	u-v2–v-v2	u-v2–v-down	u-v2–v-up
u-v2–v-v1	u-v2–v-v2	u-v2–v-down	u-v2–v-up

Figure 2.3. Outer product array [U_{ba}] (eq. 1.9) of the composite u-type and v-type spinors illustrating the structure of the outer product array of uv's.

Since the annihilation process for the unpolarized seed fermion-antifermion pair requires a sum over spin, the four columns of Fig. 2.3 become 16 columns due to the spin summation An 8 dimension fermion has four up-spin values and four down spin values.

2.3 Block Form of [U_{ba}] Array

We now turn to partitioning the [U_{ba}] and identifying the blocks of the QUeST array implied by the partition. Each column in Fig. 2.3 becomes four columns in [U_{ba}] extended by the summation over the spin of the unpolarized seed fermions. The pattern of array entries is displayed in Fig. 2.4.

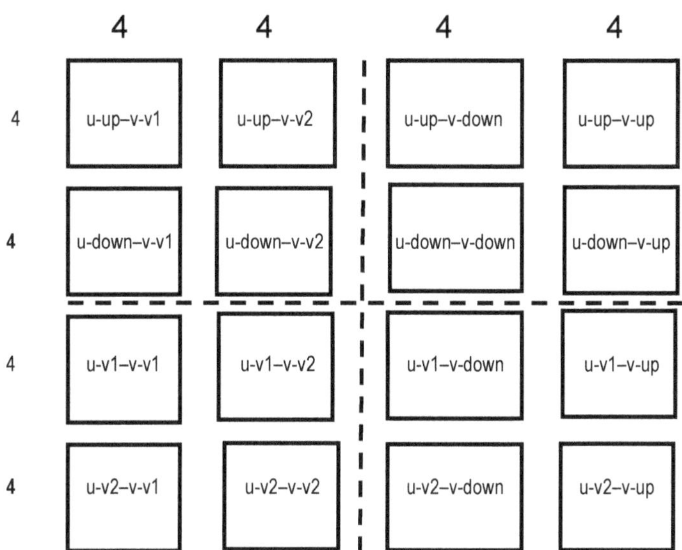

Figure 2.4. Block form of the 16 × 16 [U_{ba}] array. This is also the form of the QUeST dimension array of 256 dimensions. The blocks are divided by dashed lines that separate 64 dimension sections. These sections map to layers in QUeST (UST) as shown in Figs. C.2 and C.3.

The 4 × 4 blocks in the array match the 4 × 4 blocks that we found in earlier books such as Blaha (2020j). Figs. C.9, C.!0, and C.12 show the 4 × 4 block structure of QUeST dimensions and fermions. The sets of four blocks are the four layers of a QUeST and UST. A 64 dimension layering is evident.

We can rearrange the blocks of Fig. 2.4 to a four layer format (Fig. 2.5) similar to QUeST. Fig. 2.5 shows the block layout of Fig. 2.4 reshaped by moving the right blocks below the left blocks.

Thus we find the structure of the QUeST dimension array (Figs. C.3 and C.4) is a reflection of the structure of the seed fermion-antifermion spinor array. It has 64 sixteen dimension blocks. The blocks can be assembled in fours to give the form of each of the four layers. Each layer has 64 dimensions within it. The QUeST dimension array maps directly to our UST theory as shown in chapter 3. *BQUeST yields the QUeST dimension array structure as described in our earlier books!*

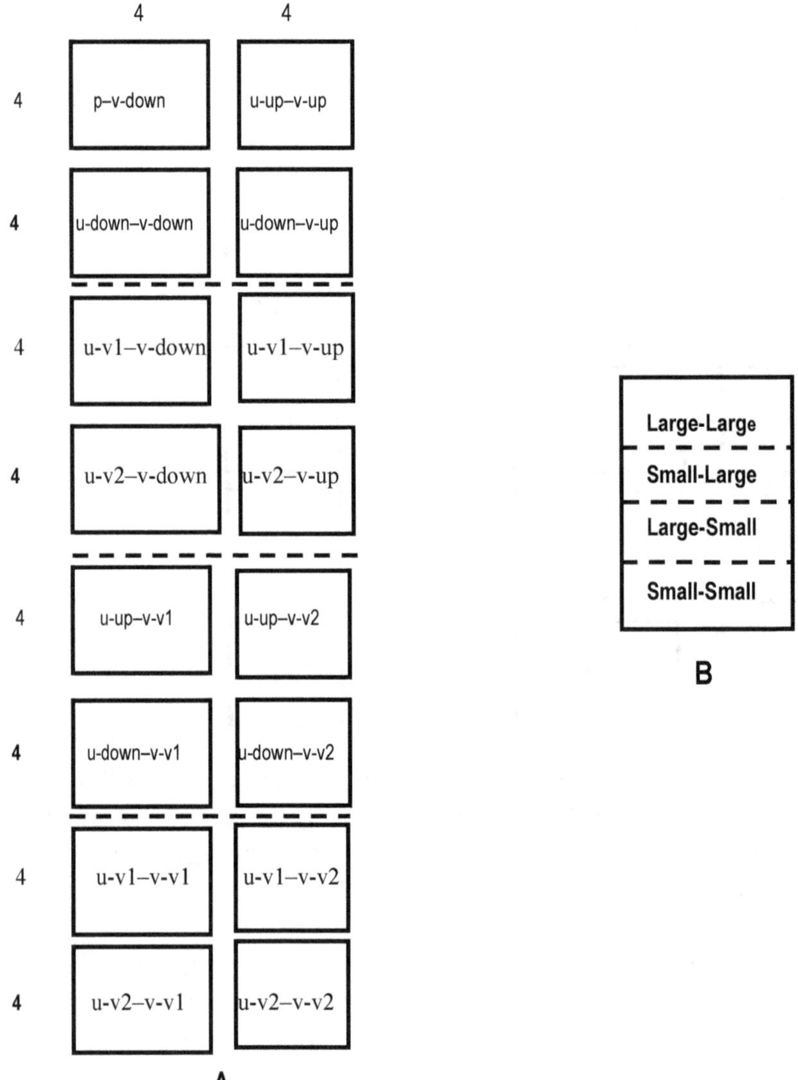

Figure 2.5. Fig. A: Block form of the 16 × 16 $[U_{ba}]$ array arranged to a 32 × 8 form to display a layer structure similar to QUeST. Fig. B: The layers were grouped by sector type and ordered from "large" spinor components to "small" spinor components.

2.4 QUeST Dimension Array Inplied by 32 × 8 Spinor Array Form

The form of the 32 × 8 spinor array of Fig. 2.5 maps directly to the form of the QUeST dimension array. Fig. 2.6 shows the four layer dimension array's form. Each • represents a dimension in this Mathematical Picture Language depiction.

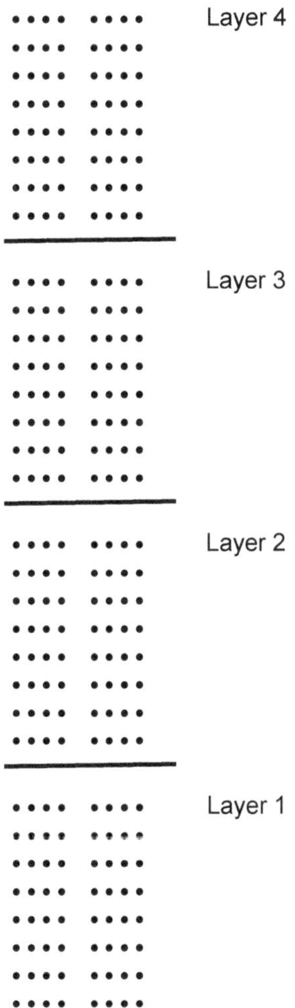

Figure 2.6. The 32 complex quaternion dimension QUeST array subdivided into 4 layers of 8 rows..Each layer will be seen to map to a block of fundamental group representations as shown in Figs. 2.7 and 2.8.

The layers of Fig. 2.6 have their dimensions allocated to various QUeST (UST) groups in Fig. 2.7. The breakdown of layers into blocks of 16 dimensions is shown in Fig. 2.8 for two out of the four layers.

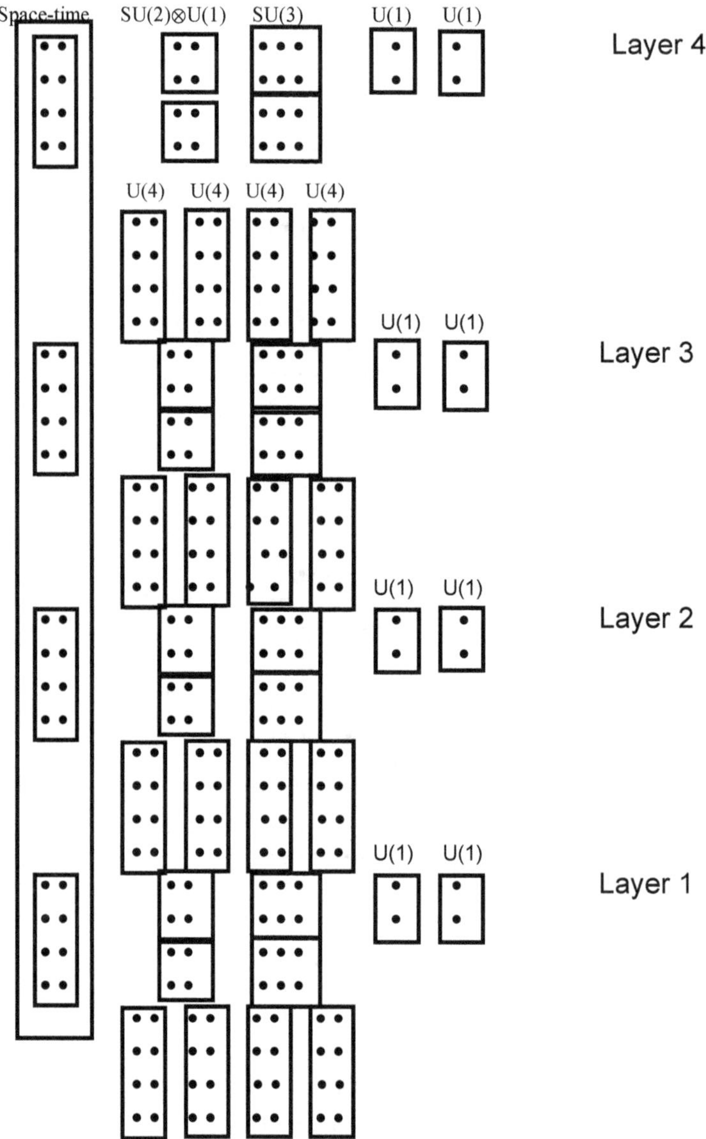

Figure 2.7. The four layers of QUeST internal symmetry groups (and space-time) for 32 dimension complex quaternion space. Note: each row has an 8 • complex quaternion. Note the left column of blocks combine to specify a 4 dimension complex quaternion space-time. Note each layer requires 64 dimensions.

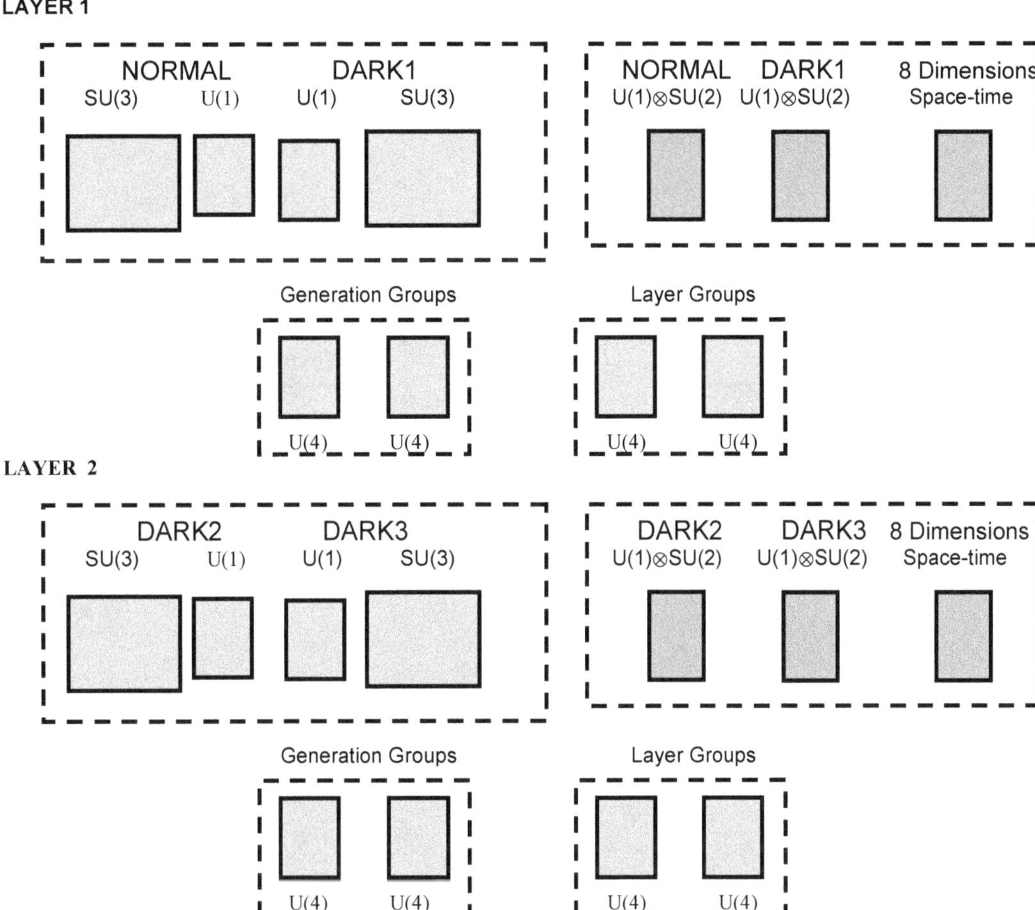

Figure 2.8. Sets of 16 dimension blocks for the first two layers of the 32 × 8 array. Each "dashed" block (regardless of its apparent size) is a 4 × 4 = 16 array of dimensions. This set of 8 blocks contains the 8×16 = 128 dimensions of layer 1 and Layer 2.

2.5 Progression from BQUeST to UST

Fig. 2.9 shows the progression from BQUeST to UST. The structure of the Unified SuperStandard Theory (UST) reflects the structure of the spinor array constructed from the seed fermions. *Thus we have support at the level of Elementary Particles for the possibility of a fundamental one fermion BQUeST level residing in the eight dimension Magaverse space-time.*

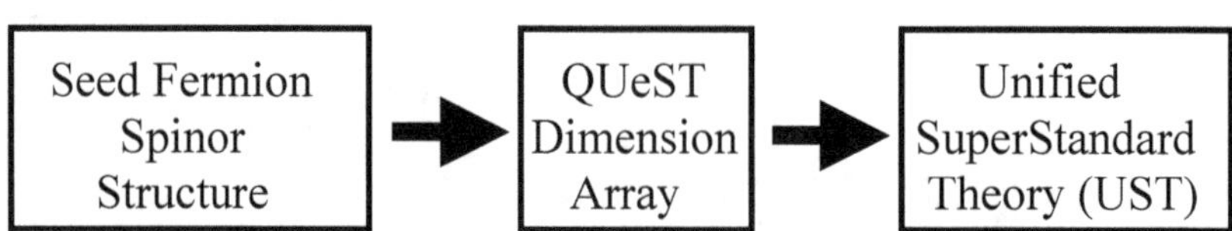

Figure 2.9. The seed fermion spinor array structure is mirrored by the QUeST dimension structure. This is evidence for the seed origin of QUeST of BQUeST.

3. QUeST Universe Structure & its Relation to UST

BQUeST, QUeST, and UST form a complete theory of elementary particles and Gravitation. In chapter 1 we derived QUeST from BQUeST. In this chapter we show that QUeST implies UST: same internal symmetries, same fermion, vector boson, and Higgs particles. *This chapter shows that starting from BQUeST with one fermion residing in an 8 dimension Megaverse one can generate QUeST and show it is consistent with UST.*

The probability of a consistent match of quaternion-based QUeST and logic-based UST would seem to be very low. The consistency suggests that we are on the right track for a complete theory of elementary particles and gravitation.

3.1 Internal Symmetries

QUeST has a 256 dimension space. This space furnishes a fundamental representation for U(128). This symmetry must be broken to obtain the internal symmetries and space-time of UST (and of the Standard Model within it). A pattern of symmetry breaking[26,27] that leads to UST and the Standard Model is:

1. U(128) broken to U(32)⊗U(32)⊗U(32)⊗U(32) giving four layers
2. Each U(32) broken to a 32 dimension U(16) giving "Normal" and "Dark" sectors
3. Each U(16) broken to a 16 dimension U(8) giving subdivisions of "Normal" and "Dark"
4. Each U(8) broken to SU(2)⊗U(1)⊗SU(3)⊗U(1)⊗U(4)⊗U(4)⊗(4 dimension space-time part)

Figure 3.1. Pattern of symmetry breaking of QUeST U(128).

This pattern results in a "tiling" of the 256 dimension QUeST array with 16 tiles composed of the groups:

SU(2)⊗U(1)⊗SU(3)⊗U(1)⊗(4 dimension space-time part) or U(4)⊗U(4) (3.1)

The space-time parts combine to give a four complex quaternion space-time from which real four dimension space-time results. One might ask how our real space-time results. One simple answer is that it is the result of our real four dimension measuring tools. We can only "see" things of our four dimensions ultimately because of conservation of momentum. Things within our space-time are visible. Things outside our space-time are

---

[26] The pattern presented here is an alternative view of the QUeST dimension array. The BQUeST derivation of Chapter 1 is fully consistent with it.

[27] Suggested by the derivation of QUeST in a two-step process: first using the one dimension fermion's four spinors (the fermion is in a 4-dimension external space) to boost 1-dimension BQUeST to a 4 × 4 array; then assuming each of the array components is a fermion using each's 4-spinors boost to obtain a 16 × 16 dimension QUeST array. See Blaha (2020i) for details.

not. Things partly within and partly without would be thought to have anomalous features (such as anomalous mass), and would usually be discarded as experimental, unreproducible errors. In the case of the universe, distortions in its structure could be attributed to objects beyond the universe.

The symmetry breaking of Fig. 3.1 lead to the internal symmetries of QUeST:[28]

$$[SU(2)\otimes U(1)\otimes SU(3)\otimes U(1)]^8\otimes U(4)^{16} \tag{3.2}$$

The corresponding layout of the QUeST dimensions appears in Fig. 3.2. There are four layers of internal symmetries in QUeST. Each layer has two SU(2)⊗U(1)⊗SU(3)⊗U(1) groups,[29] two U(4) Generation groups, two U(4) Layer groups, and two U(1) Fermion groups (described later).[30] See Figs. 3.2 and 3.3.

[28] QUeST/UST also contains the U(4) Species group, which follows from the General Relativity of the space-time.

[29] SU(2)⊗U(1)⊗SU(3)⊗U(1) does not have an SU(5) covering group. The groups in QUeST do not support SU(5) characterizations.

[30] We have replaced the Dark U(2) group seen earlier here and in previous books with a pair of U(1) groups because we were better able to achieve the $4 \times 4 = 16$ dimensions pattern that seems present.

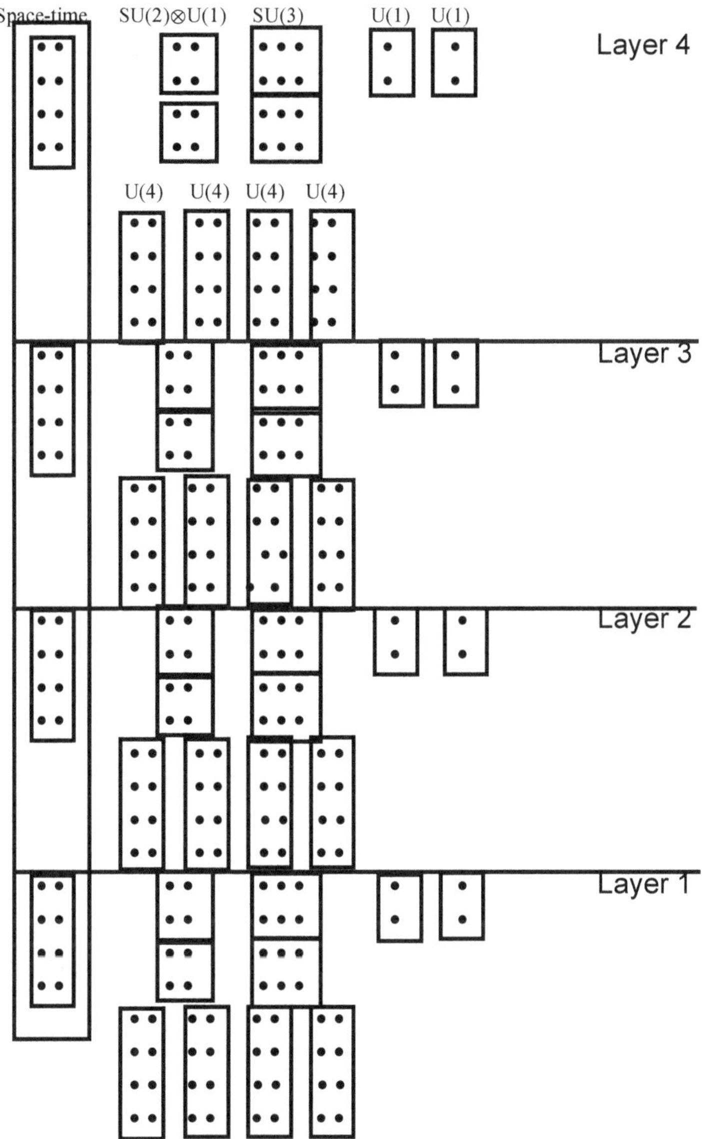

Figure 3.2. The four layers of QUeST internal symmetry groups (and space-time) for its 32 dimension complex quaternion space. This is a changed from earlier books on UST: U(2) → *U(1)⊗U(1)*. It is done to support the symmetry breakdown pattern above. Note: each row has an 8 • complex quaternion. Note the left column of blocks combine to specify a 4 dimension complex quaternion space-time. Note each layer requires 64 dimensions.

	NORMAL 4	NORMAL 4	DARK 4	DARK 4
4	SU(2)⊗U(1)⊗SU(3)⊗U(1) 4 Space-time Dimensions	Generation + Layer Groups	SU(2)⊗U(1)⊗SU(3)⊗U(1) 4 Space-time Dimensions	Generation + Layer Groups
4	SU(2)⊗U(1)⊗SU(3)⊗U(1) 4 Space-time Dimensions	Generation + Layer Groups	SU(2)⊗U(1)⊗SU(3)⊗U(1) 4 Space-time Dimensions	Generation + Layer Groups
4	SU(2)⊗U(1)⊗SU(3)⊗U(1) 4 Space-time Dimensions	Generation + Layer Groups	SU(2)⊗U(1)⊗SU(3)⊗U(1) 4 Space-time Dimensions	Generation + Layer Groups
4	SU(2)⊗U(1)⊗SU(3)⊗U(1) 4 Space-time Dimensions	Generation + Layer Groups	SU(2)⊗U(1)⊗SU(3)⊗U(1) 4 Space-time Dimensions	Generation + Layer Groups

Figure 3.3. Four layers of Internal Symmetry groups in QUeST. The groups in each layer are independent of those in other layers. The groups in each block of each layer are independent of those in the other blocks. Each block contains 16 dimensions. The block dimensions furnish fundamental representations for the groups listed. The entire set of blocks contains 256 dimensions. Each layer contains 64 dimensions. The first two columns are for the "Normal" sector. The last two columns are for the "Dark" sector (although most of the Normal sector is Dark observationally at present.) This Figure also holds for UST with the addition of Fermion groups.

Note: The columns labeled NORMAL in Figs. 3.3, 3.4 and later include three layers that are Dark in the sense that their interactions and matter are not as yet found experimentally. The columns labeled DARK are Dark, now and in the future. DARK sectors are distinct from NORMAL sectors.

3.2 QUeST → UST Symmetries

The 16 tiles of Fig. 3.3 contain various symmetry groups and space-time. They are multiple copies of the same internal symmetries. However their coupling constants, particle masses, and symmetry breaking parameters will differ in general.[31]

We are familiar with part of the one known tile: three known generations of Normal fermions, the SU(2)⊗U(1)⊗SU(3) ElectroWeak and Strong interactions, and

[31] If the Blaha calculation of the ElectroWeak and Strong coupling constants based on vacuum polarization holds, then all copies of SU(2)⊗U(1)⊗SU(3) have the same coupling constant values. See Blaha (2020c) and earlier books. IF the Blaha vacuum polarization calculation of coupling constants *in all QUeST layers* does *not* hold, then Grand Unified Theory (GUT) unification is doubtful. IF the Blaha vacuum polarization calculation of coupling constants *in all QUeST layers holds*, then Grand Unified Theory (GUT) unification becomes possible.

some masses, couplings and symmetry breaking parameters. The space-time is also known. The other tiles are completely unknown although our Quaternion-based theory suggests their likely form.

We now turn to mapping the above features of QUeST to UST. The QUeST symmetries listed in columns 1 and 3 of Fig. 3.3 do not appear in UST. We augment UST, which is derived from axioms, by including these U(1) symmetries. The space-time coordinates are separate in UST. UST has 3+1 dimensions separate from internal symmetry; QUeST unites all dimensions for internal symmetry and space-time—one of its motivating goals.

Otherwise the pattern of QUeST symmetries matches the pattern of UST. We used this correspondence to determine QUeST symmetries of columns 1 and 3 in Fig. 3.3 while adding a U(1) Fermion group (discussed below) to each of their tiles. Further we use the UST symmetries of columns 2 and 4 of Fig. 3.4 to determine the corresponding columns of QUeST. These choices are appropriate since the QUeST dimension array does not inherently determine the physical nature of QUeST groups. Instead we see a *Correspondence Principle* where UST determines the physics of the QUeST array. Consistency of the QUeST array groups with the UST groups is required. This section demonstrates consistency if Fermion U(1) groups are added to UST. Satisfyingly, the U(1) groups have a natural physical interpretation in UST.

	NORMAL		DARK	
4	SU(2)⊗U(1)⊗SU(3)	Generation + Layer Groups	SU(2)⊗U(1)⊗SU(3)	Generation + Layer Groups
4	SU(2)⊗U(1)⊗SU(3)	Generation + Layer Groups	SU(2)⊗U(1)⊗SU(3)	Generation + Layer Groups
4	SU(2)⊗U(1)⊗SU(3)	Generation + Layer Groups	SU(2)⊗U(1)⊗SU(3)	Generation + Layer Groups
4	SU(2)⊗U(1)⊗SU(3)	Generation + Layer Groups	SU(2)⊗U(1)⊗SU(3)	Generation + Layer Groups

Figure 3.4. Four layers of Internal Symmetry groups in UST from Blaha (2020c) and earlier books such as Blaha (2018e). The groups in each layer are independent of those in other layers. The groups in each block of each layer are independent of those in the other blocks. The first two columns are for the "Normal" sector. The last two columns are for the "Dark" sector (although most of the Normal sector is Dark observationally at present.) Note columns 1 and 3 of UST omit a U(1) group relative to Fig. 3.3. Also space-time is separate and not included with of the set of symmetries in UST. The UST tiles in columns 1

and 3 have 10 dimensions (space-time dimensions omitted also); the tiles in columns 2 and 4 have 16 dimensions.

We now consider the Generation, Layer and Fermion groups. These groups occur in UST. We map them back to QUeST using the previously mentioned Correspondence Principle..

3.3 U(4) Generation, U(4) Layer and U(1) Fermion Groups

The Generation, and Layer groups were introduced in UST. (See Blaha (2018e), 2020c) and earlier books.) Their basis is in number conservation laws such as Baryon conservation. The Fermion group, now introduced, is also based on conservation laws:

We now describe the Generation and Layer groups briefly. Blaha (2020c) provides a more detailed description as does earlier books. The Fermion group is described later.

3.4 The Generation Group

We define two particle number operators for normal up-quark particles and down-quark particles, B_{uq} and B_{dq}. Similarly we define two particle number operators for normal species "e" (electron) particles and species "ν" particles, B_e and B_ν. Similarly we define Dark matter equivalents:[32] B_{De}, $B_{D\nu}$, B_{Duq}, and B_{Ddq}.

In the absence of interactions these fermion particle number operators are conserved. Each set are "diagonal" operators within a U(4) group. Thus we have a normal U(4) Generation Group and a Dark U(4) Generation group *for each layer.*.

On this basis we find there are four generations of each species in the normal and in the Dark matter sectors since U(4) groups have 4-dimension fundamental representations. One generation of normal fermions with large masses has not as yet been found in the lowest layer; three generations of normal fermions have been found. .

The broken symmetry gauge vector bosons of the Generation Group also have large masses. If the conservation of the fermion particle numbers is broken then we view it as a consequence of Generation Group symmetry breaking.

The Generation group of each layer generates interactions between the fermions of that layer. See Fig. C.5 of Appendix C for the interaction between fermions generated by the Generation group of layer 4.

3.5 The Layer Group

The set of particle number operators can be extended if we take account of the fourfold fermion generations. To further refine the set of particle number operators we temporarily neglect all interactions that would violate conservation laws for the set.

We therefore subdivide the above particle number set into four particle numbers per generation. For the ith generation we define

---

[32] By analogy, we assume that there are four species of Dark matter: charged Dark leptons, neutral Dark leptons, Dark up-type quarks, and Dark down-type quarks. Thus we are led to the Dark particle numbers: Dark Baryon Numbers, and Dark Lepton Numbers shown above.

L_{ie} – The "e" species particle number for the ith generation
L_{iv} – The ν species particle number for the ith generation
L_{iuq} – The up-quark species particle number for the ith generation
L_{idq} – The down-quark species particle number for the ith generation

L_{iDe} – The Dark "e" species particle number for the ith generation
L_{iDv} – The Dark ν species particle number for the ith generation
L_{iDuq} – The Dark up-quark species particle number for the ith generation
L_{iDdq} – Dark down-quark species particle number for the ith generation

for each generation i = 1, 2, 3, 4. Individual fermions have positive $L_{ia} = +1$ values and anti-fermions have negative $L_{ia} = -1$ values for species a = 1, 2, 3, 4 (with the three color subspecies of quarks treated as part of one species.)

At this point we have four particle number operators for each generation. We define a group framework for each set of particle numbers. The simplest way is to assume that each generation consists of the four layers with the particles in each generation in a U(4) fundamental representation. Then each generation has a U(4) Layer group with the generation's four number operators (above) as its diagonal operators. We call this group the Layer Group of the ith generation L_{ia}. With four generations we obtain four U(4) Layer groups for normal matter. In addition there are four U(4) Dark Layer groups. See Fig. 3.3.

The consequence of this expansion of particle numbers and groups is that the set of fermions increases fourfold. We now have four layers, with each having four generations. Experimentally, we know of three generations of fermions—the lowest generations of the lowest level. The remaining generation and the three additional levels of fermions are of much higher mass and yet to be found.

See Blaha (2020c) and (2018e) for a detailed discussion of the Layer Groups. We note in passing that the symmetries of these number operators are badly broken. Yet the underlying group structure remains.

Note each Layer group provides an interaction among layers: the ith Layer group provides an interaction between the four i^{th} generations. See Fig. C.5 for the four Layer groups' interactions between fermions for the four generations. Layer groups provide the only interactions between layers.

3.6 The Fermion Groups

For the NORMAL sector of each layer there is a U(1) Fermion group.[33] There is a similar U(1) Fermion group for the DARK sector of each layer. See Fig. 3.3. These groups were introduced in QUeST. We have added them to UST since they have natural roles.

---

[33] Earlier versions of QUeST in this year had a broken U(2) Dark symmetry group that gave interactions between Normal and Dark particles. Here we replace this group with U(1)⊗U(1) Fermion groups due to a tile by tile conserved fermion number. Appendix C uses the Dark U(2) group. The conent of that appendix can be modified to Fermion groups by replacing each Dark U(2) with U(1)⊗U(1).

We will attribute them to fermion numbers conservation[34] for each layer and for NORMAL and DARK separately—one conservation law for total fermion number in each tile of columns 1 and 3 of Fig. 3.3. *In each of the columns 1 and 3 tiles fermion number equals the sum of Baryon number and Lepton number. It appears to be conserved in particle interactions. It can be conserved even if Baryon number conservation and Lepton number conservation were violated. The sum of all NORMAL Fermion numbers F_{tot} is also conserved. The sum of all DARK Fermion numbers F_{Dtot} is conserved as well.*

For the NORMAL and DARK sectors of each layer of fermions we define an additive Fermion number N_f that equals the number of fermions of that layer minus the number of antifermions of that layer. In an interaction where fermion number is conserved, the input N_f equals the output N_f. This conservation law, which resembles the baryon number conservation law, is broken by Layer group interactions. Thus the broken Fermion groups, like the Generation and Layer groups, are based on a conserved number. The gauge fields of the Fermion groups will be denoted Y^{μ} and $Y_D{}^{\mu}$ for the NORMAL and DARK sectors of each layer respectively.

UST with Fermion groups added follows from QUeST.

Exercise: Replace the Dark U(2) group in Appendix C with U(1)⊗U(1) Fermion groups.

3.7 Particle Spectrums of QUeST: Fermions, and Vector Bosons

In this section we show that QUeST and UST have the same vector boson and fermion particle spectrums.

3.7.1 Vector Bosons

An examination of Figs. 3.2 and 3.3, which now apply to both QUeST and UST, directly yields the spectrum of gauge vector bosons. This spectrum includes the known vector bosons of the Standard Model. The list of vector boson groups and interactions *for each of the four layers* is:

<u>NORMAL Gauge Groups</u>
SU(2)⊗U(1)⊗SU(3)⊗U(1)
Generation Group U(4)
Layer Group U(4)
<u>DARK Gauge Groups</u>
SU(2)⊗U(1)⊗SU(3)⊗U(1)
Generation Group U(4)
Layer Group U(4)

3.7.2 Fermions

Having the spectrum of gauge vector bosons the fundamental fermion spectrum also follows directly from the dimensions of the fundamental group representations. See Fig. C.5 for the QUeST and UST fermion spectrums.

The basis for the fermion spectrum is twofold:

---

[34] Fermion number conservation differs from charge conservation since it applies to both charged and uncharged fermions.

1. There are four species of fermions due to the boosts of the Complex Lorentz group as shown in Blaha (2020c) and earlier books. These species are charged lepton, neutral lepton, up-type quark, and down-type quark.[35]

2. The leptons were seen to be singlets. The quarks were seen to be Color triplets.

3. With these points in mind there are 8 fermions for each of the eight appearances of SU(2)⊗U(1)⊗SU(3)⊗U(1) in Fig. 3.3. See Fig. C.5.

3.8 Other QUeST-UST Topics

There are a number of important UST topics that are presented in Blaha (2020c) and (2018e) as well as other book by the author. They include:

1. Color Confinement
2. Gravitation Potential at Earthly, Galactic, and intergalactic Distances
3. Higgs Mechanism and Symmetry Breaking. Fermion and Gauge Boson Masses
4. Monads From factorization of Wave Functions to Eliminate Quantum Entanglement
5. Determination of α and Other Coupling Constants
6. Evidence for Faster-than-Light Particles and their Implications
7. Complex Gravitation and the Species group
8. An Equipartition Principle for Universe Matter and Energy Abundance

QUeST-UST is a complete theory of the framework of elementary particle phenomena.

---

[35] While this was shown as early as 2007 by the author, it became very evident as a general form when it was shown to hold in the UTMOST Megaverse earlier in this year.

4. The BMOST Derivation of the UTMOST Megaverse

This chapter describes in detail the possible birth of the 1024 dimension UTMOST Megaverse from a fermion-antifermion annihilation in a separate 10 dimension space.[36] It describes the BMOST theory developed in 2020 in earlier books by the author.

BMOST[37] assumes the Megaverse begins from the annihilation of a fermion-antifermion pair in a 10 dimension space-time.

The derivation of the 32 × 32 = 1024 dimension UTMOST array requires a seed fermion, which we call the *urfermion*,[38] to be a 10 dimension fermion. The 32 spinor components of the 10 dimension urfermion will be used to generate the 32 × 32 dimension array of UTMOST.

We propose a picture of a 10 dimension fermion —the urfermion—that acts as the seed of the consequent Megaverse. The evolution (perhaps instantaneous) of the Megaverse begins with seed-antiseed annihilation at great energy to become the 1024 dimension UTMOST Megaverse. *The urfermion, which is off shell, has 32 independent spinor components that leads to the 1024 dimensions of the UTMOST Megaverse.* It generates internal symmetries and the space-time of UTMOST. Subsequently universes appear within the Megaverse as described in chapter 1.

The derivation parallels that of universe birth in the eight dimension Megaverse. Appendix E describes features of the UTMOST space.

4.1 BQUeST Origin

We begin with the assumption that the UTMOST Megaverse originates in a 10 dimension space..[39] The 10 dimension seed urfermion has 32 spinor components. (There are 1024 Dirac matrices, denoted G_{ab} for a, b = 1, …, 32, and ten 32×32 Dirac γ matrices.)

We view the 10 dimension space as the location of the origin of the Megaverse. We picture an off shell seed urfermion and an off shell seed antiurfermion as appearing as fluctuations in the 10 dimension space vacuum. They "annihilate" to create a

[36] The prevalence of 10 dimension spaces in SuperString theories raises the possibility of a relation of the Megaverse to a SuperString space.

[37] See Blaha (2020d) and later books.

[38] We use the Germanic prefix ur- to signify original (or earliest). The author used this prefix in the 1970s in published Physics Letters B papers on quarks and leptons including discrete scaling fermion masses.

[39] The UTMOST Megaverse is described in Appendix E.

Megaverse particle.[40] The Megaverse particle expands to become the Megaverse—the home of universes.

We will show the 32 independent spinor components of the 10 dimension off shell seed urfermion and of the off shell anti-seed urfermion generate the 32 × 32 dimension array of QUeST.[41]

4.2 A Model Dynamics of the Seed – Anti-Seed Generation of the Megaverse

The seed urfermion must have a dynamics that enables it to generate the 1024 dimension array. In this section we describe a model dynamics for the annihilation of a seed–antiseed state into a scalar Megaverse particle. The model must have certain features that accomplish the goal:

1. Since the target dimension array is not symmetric the PseudoQuantum formulation[42] of Quantum Field Theory will be seen to be required.

2. Since the quantum field theory is in 10 dimensions we must use Two-Tier coordinates,[43] which eliminate all divergences in perturbation theory using exponential damping of all integrations.

These extensions of quantum field theory were developed by the author in the 1970's and the early 2000's. They are both described in Appendix A together with references to earlier papers and work.

We define a model lagrangian for the seed urfermion sector. The seed urfermion is represented by *two* quantum fields[44] ψ_1 and ψ_2. The scalar universe particle has two quantum fields A_{1ab} and A_{2ab} with indices a and b.

The lagrangian terms are

$$\mathscr{L} = F^{1\mu\nu} F^2_{\ \mu\nu} + \overline{\psi}_2 (i\gamma\cdot\partial - m)\psi_1 + \overline{\psi}_1 (i\gamma\cdot\partial - m)\psi_2 - e_0\overline{\psi}_2 A_2\psi_1 - \overline{\psi}_1 A_1\psi_2 \tag{4.1}$$

where m is the mass of the seed urfermion, and where $A_i = A_{iab}G_{ab}$ (summed over a and b) are the Megaverse fields.

---

[40] Appendix D provides experimental and theoretical evidence for a particle view of universes. Comments by DeWitt and others on quantum universes support this view. A quantum Megaverse (required for consistency) must be aparticle too.

[41] The 10 dimension space may be a 10 complex octonion space-time. We choose to extract the real coordinate of each complex octonion 16-tuple to construct a real 10 dimension space-time within which seed urfermion dynamics occurs. This is analogous to the case of complex 4 dimension space-time being restricted to real complex space-time in conventional quantum field theory. Streater (2000) maintains complex space-time is needed for axiomatic quantum field theory.

[42] S. Blaha, Il Nuovo Cimento **49A**, 35 (1979) and Appendix A.

[43] See Appendix A.

[44] We use a PseudoQuantum Electromagnetic-like pair of fields also. The electromagnetic-like particle is a model of the universe. See eqs. 61 – 90 in Appendix 1-C of Blaha (2020j). The Two-Tier formulation eliminates all perturbation theory divergences in four dimensions including the notorious fermion triangle divergence. It also eliminates all perturbation theory divergences in eight dimensions.

To insure finite perturbation series results (in 10 dimensions) the fields are all functions of Two-Tier coordinates X having a vector quantum field $Y^\mu(y)$:

$$X^\mu(y) = y^\mu + i\, Y^\mu(y)/M_c^2 \tag{4.2}$$

with the Y^μ index $\mu = 0, 1, \ldots, 9$, and where M_c is an extremely large mass.

4.2.1 U(32) Species Group Symmetry for Dirac Matrix Indices

The lagrangian terms (plus additional universe terms) are invariant under a U(32) Species symmetry group. Appendix B describes the analogous U(4) Species[45] symmetry for four dimension fermions that we found for UST.

Appendix B shows the form of the (vector field) trial lagrangian for establishing the Species group of UST in conventional quantum field theory:

$$\bar{\psi}'(x)[i\gamma_\mu'(x)(\partial/\partial x_\mu - ig_8 A'_{Sk}{}^{\mu}(x)G_{Sk}) - m]\psi'(x) \tag{51.2}$$

where the matrices G_{Sk} are the 16 Dirac matrices of 4 dimension space, where k ranges from 1 through 16, and where the $A'_{Sk}{}^{\mu}$ fields.are 16 gauge fields for the Species group.

In developing the U(4) Species group symmetry for four dimension UST we related the indices of the Dirac matrices $[G_{Sk}]_{ab}$ where a, b = 1, 2, 3, 4 to the space-time coordinates using vierbein transformations. In this regard Appendix B states:

> A complex transformation of types II and III in Fig. 50.1 has the form:
>
> $$U(x'')^{\mu}{}_{\nu} = w^{\mu}{}_{a}(x'')[\exp(i \textstyle\sum_k \Phi_k(x'')\tau_k)]^{a}{}_{b}\, 1^{b}{}_{\nu}(x'')$$
> $$U^{-1}(x'')^{\mu}{}_{\nu} = w^{\mu}{}_{a}(x'')[\exp(-i \textstyle\sum_k \Phi_k(x'')\tau_k)]^{a}{}_{b}\, 1^{b}{}_{\nu}(x'')$$
>
> where τ_k is a U(4) generator matrix. Its infinitesimal transformation is approximately
>
> $$U(x'')^{\nu}{}_{\beta} \approx \delta^{\nu}{}_{\beta} + i \textstyle\sum_k \Phi_k(x'')[\tau_k]^{\nu}{}_{\beta} \tag{50.12}$$
> $$U^{-1}(x'')^{\nu}{}_{\beta} \approx \delta^{\nu}{}_{\beta} - i \textstyle\sum_k \Phi_k(x'')[\tau_k]^{\nu}{}_{\beta}$$
>
> using the *vierbein* flat space-time limits
>
> $$w^{\mu}{}_{a}(x'') \approx \delta^{\mu}{}_{a} \tag{50.12a}$$
> $$1^{b}{}_{\nu}(x'') \approx \delta^{b}{}_{\nu}$$
>
> where
>
> $$\Phi_k(x) = \int^{x} dy_\lambda\, A_{Rk}{}^{\lambda}(y) \tag{50.13}$$
>
> Then

[45] The U(4) Species symmetry group is part of UST. It originates in the Complex General Reloativity of four dimension coordinates as shown in Appendix A and Blaha (2020c) as well as earlier books.

$$\Gamma_{R\ \lambda\mu}^{\ \sigma} = -\tfrac{1}{2}i\{\sum_k A_{Rk}(x'')_\mu[\tau_k]^{\sigma}{}_{\lambda} + \sum_k A_{Rk}(x'')_\lambda[\tau_k]^{\sigma}{}_{\mu}\} \quad (50.14)$$
$$= A_R{}^{\sigma}{}_{\mu\lambda} + A_R{}^{\sigma}{}_{\lambda\mu}$$

(summed over k) with the matrix $A_R{}^{\sigma}{}_{\mu\lambda}$ given by

$$A_R{}^{\sigma}{}_{\mu\lambda} = -\tfrac{1}{2}i\sum_k A_{Rk\mu}[\tau_k]^{\sigma}{}_{\lambda} \quad (50.15)$$

with $A_R{}^{\sigma}{}_{\mu\lambda}$ transformable to matrix row and column numbers

$$A_{R_{flat}}{}^{\mu a}{}_{b} = A_{R_{flatk}}{}^{\mu}[\tau_k]^{\sigma}{}_{\lambda}\delta_{\sigma}{}^{a}\delta^{\lambda}{}_{b} \quad (4.3)$$

using the flat space-time vierbein values, and so $A_{R_{flat}}{}^{a}{}_{\mu b}$ may be written in matrix form as

$$A_{R_{flat}\mu} = -\tfrac{1}{2}i\sum_k A_{R_{flatk}\mu}\tau_k \quad (50.16)$$

In the flat space-time limit the $A_{Rk}{}^{\lambda}(y)$ becomes the Coordinate Species group U(4) gauge fields $A_{R_{flatk}}{}^{\lambda}(y)$.

The above extract shows that space-time dimensions can be directly related to matrix indices. We will use this feature to map the indices of the Megaverse fields A_{iab} to internal dimensions of the Megaverse particle. The Megaverse dimensions, which number 1024 in the 10 dimension space-time, become the 1024 dimensions in UTMOST. (It should be remembered that the UTMOST Megaverse has 1024 dimensions —much more than the 256 dimensions of QUeST.

4.3 The Interaction Generating a Megaverse Particle

The diagram for the creation process is in Fig. 4.1. A seed urfermion and an antiseed urfermion annihilate to create a Megaverse particle.

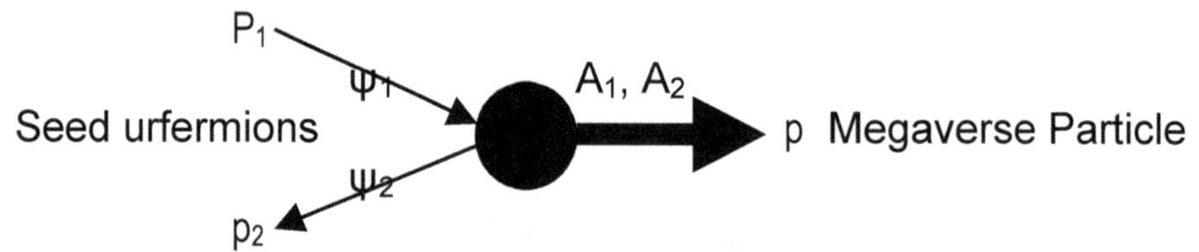

Figure 4.1. Diagram for the transition of seed - antiseed urfermions to a Megaverse particle.

The lowest order S-matrix element for this interaction has the form:

$$S = \int d^8x\, [\overline{\psi}_2 A_2 \psi_1 + \overline{\psi}_1 A_1 \psi_2] \quad (4.4)$$

After evaluating the integration we find S has term of the form

$$\overline{v}_{2a}(p_2)\ A_{1ab}(p)\ u_{1b}(p_1) \tag{4.5}$$

Rearranging the expression we obtain the trace expression

$$\mathrm{Tr}\ \ u_{1b}(p_1)\overline{v}_{2a}\ (p_2)\ A_{1ab}(p) = U_{ba}A_{1ab}(p) \tag{4.6}$$

where U_{ba} is an array composed of u and $\overline{v}$.

Note the interaction can only take place for far off shell seed and antiseed fermions. Then the 32 spinor components of each fermion are independent and can be treated as independent.

Megaverse production (Fig. 4.1) has momentum conservation due to the off shell urfermions. The diagram has two interesting cases:

1. If A_{1ab} and A_{2ab} are both proportional to a Kronecker δ_{ab} then the S matrix expression of eq. 4.4 simplifies to the scalar A_i case.
2. If A_{1ab} and A_{2ab} are not proportional to a Kronecker δ_{ab} then the S matrix expression can be expressed as a sum of terms for each value of a and b:

$$S = \Sigma\ S_{ab} \tag{4.7}$$

where

$$S_{ab} = \int d^8x\ [\overline{\psi}_{2a}A_{2ab}\psi_{1b} + \overline{\psi}_{1a}A_{1ab}\psi_{2b}] \tag{4.8}$$

The second expression eq. 4.8 suggests each "ab" pair of values marks a channel from the urfermions to the Megaverse particle. Thus we can view the Megaverse particle as receiving 1024 "streams" from the fermion-antifermion pair.

In section 4.4 below we suggest the streams "amalgamate" to form a 1024 dimension Megaverse.

4.4 U(16) Symmetry of A_i

The lagrangian in eq. 4.1 has a U(32) symmetry that is analogous to the 4 dimension U(4) Species group described in section 51.1 of Appendix B. This U(32) symmetry exists in spinor space. The matrix, $[U_{ba}]$ with its 32-spinor indices a and b, has a symmetry under the U(32) transformation:

$$U[U_{ba}]U^{-1} \tag{4.9}$$

U(32) symmetry requires the A_i factors in eqs. 4.5 and 4.6 to also have a U(32) transformation of the form:

$$A_i' = UA_iU^{-1} \tag{4.9a}$$

This symmetry will be seen to impact on the form of the UTMOST dimension array.in chapter 5.

4.5 Indices of [A_{1ab}] Transformed to Internal Dimensions

Eq. 4.5 shows the indices of the array, [A_{1ab}], have 32 · 32 = 1024 values due to the 32×32 independent values of the products: $u_{1b}(p_1)\overline{v}_{2a}(p_2)$. Eq. 4.3 above shows that indices can be transformed to coordinates.

Therefore we can take the array of indices [ab] and transform them to coordinates using a 10-space vierbein.[46] We can take the coordinates that emerge as each corresponding to a one dimension line (a stream?) within the Megaverse particle. Subsequently, or simultaneously, after creation the energy-momentum within the streams it can be distributed/flow into the 8 dimension space-time of the UTMOST Megaverse. The internal energy-momentum must sum to zero for momentum conservation. So there will be patches of positive and negative energies. (Negative energies may be absorbed by a vacuum rearrangement.)

The independence of the spinor components of each of the two seed urfermion fields guarantees the elements of the array of indices form an array of independent dimensions.

4.6 The Difference between Indices and Internal Dimensions

The array [A_{1ab}] is an array with indices. We now consider the relation of indices and internal dimensions of an entity. First we consider the case of a cube, which can be viewed as an entity with three indices. Its external indices can be taken to label the length, width and height of the cube. Within the cube we can establish a coordinate system with three dimensions labeling points within it. Thus external indices can be viewed as mapping to internal dimensions within.

Perhaps a better example is a Black Hole. It exists in 4-dimension space. Within its event horizon the coordinates of the external space can be continued. It is also possible to replace them with an internal coordinate system of four dimensions. This possibility could be supported by the role of radial dimension externally becoming effectively a time coordinate internally.

We thus conclude the indices of A_{iab} can be viewed as specifying coordinates (*dimensions)* within A_i. We have mapped the seed urfermion to a 32 × 32 = 1024 dimension array. This array serves as the dimensions of the UTMOST Megaverse space.

4.7 Origin of the Megaverse

We can then envision the possibility that the Megaverse (multiverse) started by the annihilation of a seed urfermion and antiurfermion in a 10-dimension space. It "acquired" the 1024 dimensions of the UTMOST Megaverse to form a space that includes the 7+1 dimension space-time at the Megaverse creation point as well as

[46] The Megaverse undoubtedly has an 8-dimension General Relativity. For small neighborhoods of a point the vierbein becomes the identity as eq. 50.12a above shows.

internal symmetries. This process could occur, as it likely does, instantaneously. The seed urfermion-antirfermion are necessarily in a 10-dimension space as we showed in previous books

4.8 Nature of the Seed Urfermion and Antiurfermion

The type of the seed urfermion and antiurfermion in 10-dimension space can be partly ascertained. In Blaha (2020e) we showed that QUeST and UTMOST fermions occur in four species: charged leptons, neutral leptons, up-type quarks, and down-type quarks.[47] Examples are e, ν_e, u, and d.

Similarly, there are four layers of urfermions in 10-dimension space if it supports a 10-dimension Special Relativity (and General Relativity). Each layers has sets of four generations of four species. A reasonable possibility is the seed urfermion is an *electron-type fermion* with the lowest mass of all 10-dimension space urfermions.

Thus the nature of the fundamental urfermion may be partly known.

[47] See section 1.9 for its demonstration.

5. Connection of Seed UrFermion Structure to UTMOST Dimensions Structure

Decisive experimental evidence for the origin of the Megaverse in seed urfermion-antiurfermion annihilation may never be found. There is a possibility that theoretical support for our proposed origin of the Megaverse may arise. It is based on the possibility that the structure of the UTMOST dimension array may have the "imprint" of the form of the seed urfermion-antiurfermion prior to annihilation.

The only reasonable long-term hope is that the seed urfermion-antiurfermion annihilation that produces an UTMOST Megaverse has features that appear to impress themselves on Megaverse structure. This chapter explores that possibility.[48]

5.1 Structure of 32-Spinor u and v

The thirty-two 32-spinors of a 10 dimension urfermion have the general form in Fig. 5.1.

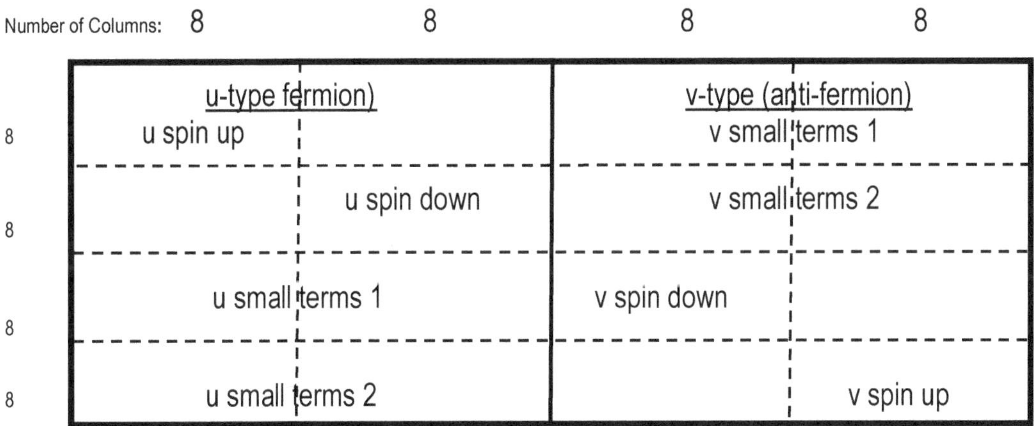

Figure 5.1. The 32 spinors of a 10 dimension spinor. Each spinor has 32 rows.

5.2 Compression of u and v into composite spinors

We can combine the sixteen u-type spinors into a composite spinor to exhibit their structure. Similarly, we can combine the sixteen v-type spinors into a composite spinor. Fig. 5.2 shows the forms of the *composite* u-type and v-type spinors.

---

48 This chapter parallels chapter 2.

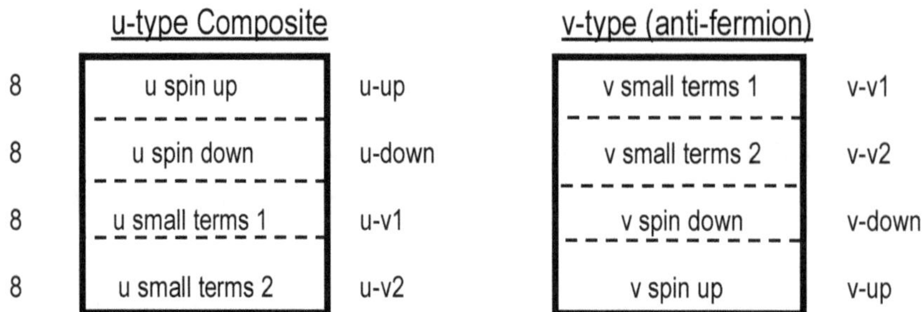

Figure 5.2. Composite u-type and v-type spinors illustrating the structure of each type of spinor. Each spinor is subdivided into 4 sections of 8 components. The third and fifth columns are symbols for each sector.

The outer product of spinor factors in eq. 4.6 is $u\bar{v}$. Fig. 5.3 contains the form implied by the composite spinors.

Since the annihilation process for the unpolarized seed urfermion-antiurfermion pair requires a sum over spin, *the four columns in Fig. 5.3 become 32 columns due to the spin summation* A 10 dimension fermion has sixteen up-spin values and sixteen down spin values.

u-up–v-v1	u-up–v-v2	u-up–v-down	u-up–v-up
u-up–v-v1	u-up–v-v2	u-up–v-down	u-up–v-up
u-up–v-v1	u-up–v-v2	u-up–v-down	u-up–v-up
u-up–v-v1	u-up–v-v2	u-up–v-down	u-up–v-up
u-up–v-v1	u-up–v-v2	u-up–v-down	u-up–v-up
u-up–v-v1	u-up–v-v2	u-up–v-down	u-up–v-up
u-up–v-v1	u-up–v-v2	u-up–v-down	u-up–v-up
u-up–v-v1	u-up–v-v2	u-up–v-down	u-up–v-up
u-down–v-v1	u-down–v-v2	u-down–v-down	u-down–v-up
u-down–v-v1	u-down–v-v2	u-down–v-down	u-down–v-up
u-down–v-v1	u-down–v-v2	u-down–v-down	u-down–v-up
u-down–v-v1	u-down–v-v2	u-down–v-down	u-down–v-up
u-down–v-v1	u-down–v-v2	u-down–v-down	u-down–v-up
u-down–v-v1	u-down–v-v2	u-down–v-down	u-down–v-up
u-down–v-v1	u-down–v-v2	u-down–v-down	u-down–v-up
u-down–v-v1	u-down–v-v2	u-down–v-down	u-down–v-up
u-v1–v-v1	u-v1–v-v2	u-v1–v-down	u-v1–v-up
u-v1–v-v1	u-v1–v-v2	u-v1–v-down	u-v1–v-up
u-v1–v-v1	u-v1–v-v2	u-v1–v-down	u-v1–v-up
u-v1–v-v1	u-v1–v-v2	u-v1–v-down	u-v1–v-up
u-v1–v-v1	u-v1–v-v2	u-v1–v-down	u-v1–v-up
u-v1–v-v1	u-v1–v-v2	u-v1–v-down	u-v1–v-up
u-v1–v-v1	u-v1–v-v2	u-v1–v-down	u-v1–v-up
u-v1–v-v1	u-v1–v-v2	u-v1–v-down	u-v1–v-up
u-v2–v-v1	u-v2–v-v2	u-v2–v-down	u-v2–v-up
u-v2–v-v1	u-v2–v-v2	u-v2–v-down	u-v2–v-up
u-v2–v-v1	u-v2–v-v2	u-v2–v-down	u-v2–v-up
u-v2–v-v1	u-v2–v-v2	u-v2–v-down	u-v2–v-up
u-v2–v-v1	u-v2–v-v2	u-v2–v-down	u-v2–v-up
u-v2–v-v1	u-v2–v-v2	u-v2–v-down	u-v2–v-up
u-v2–v-v1	u-v2–v-v2	u-v2–v-down	u-v2–v-up
u-v2–v-v1	u-v2–v-v2	u-v2–v-down	u-v2–v-up

Figure 5.3. Outer 32 × 32 product array [U_{ba}] (eq. 4.9) of the composite u-type and v-type spinors illustrating the structure of the outer product array of uv's. Each column represents 8 spinor columns, making 32 columns in all.

5.3 Block Form of [U_{ba}] Array

We now turn to partitioning the [U_{ba}] and identifying the blocks of the UTMOST array implied by the partition. Each column in Fig. 5.3 becomes eight columns in [U_{ba}] extended by the summation over the spin of the unpolarized seed fermions. The pattern of array entries is displayed in Fig. 5.4.

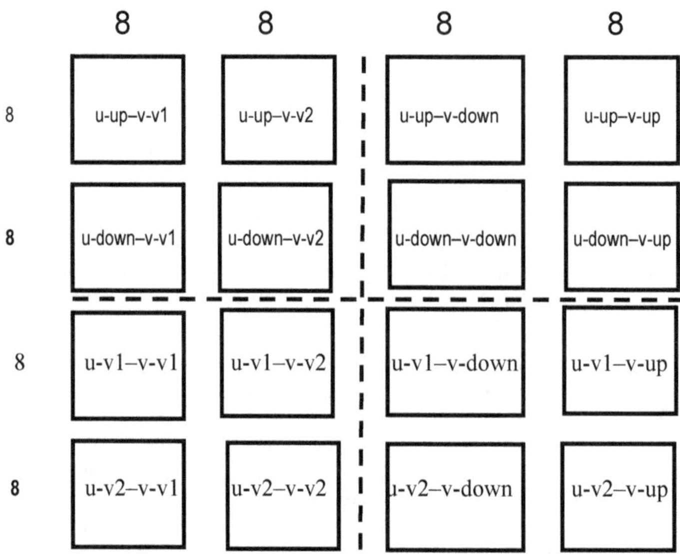

Figure 5.4. Block form of the 32 × 32 [U_{ba}] array. This is also the form of the UTMOST dimension array of 1024 dimensions. The sets of four blocks are divided by dashed lines that specify 256 dimension sections. These sections map to layers in UTMOST as shown in Figs. E.3, E.4 and E.5 for one of the four UTMOST layers. Each UTMOST layer is equivalent to a QUeST 256 dimension array.

The 8 × 8 blocks in the array match the 8 × 8 blocks (which further decompose into 4 × 4 blocks) that we found in earlier books such as Blaha (2020j). Figs. E.5 and E.7 show the 8 × 8 block structure of UTMOST dimensions *and* fermions. Each set of four blocks represents one of the four layers of UTMOST.

Fig. 5.5 shows the layout of groups in a 16 × 16 four 8 × 8 block section in Fig. 5.4.

The 8 × 8 = 64 dimension blocks show the correspondence of the seed urfermion-antiurfermion spinor structure to the UTMOST dimension structure. Given the structure of the Megaverse one could verify the possible origin of the Megaverse in BMOST urfermions in 10 space-time dimensions.

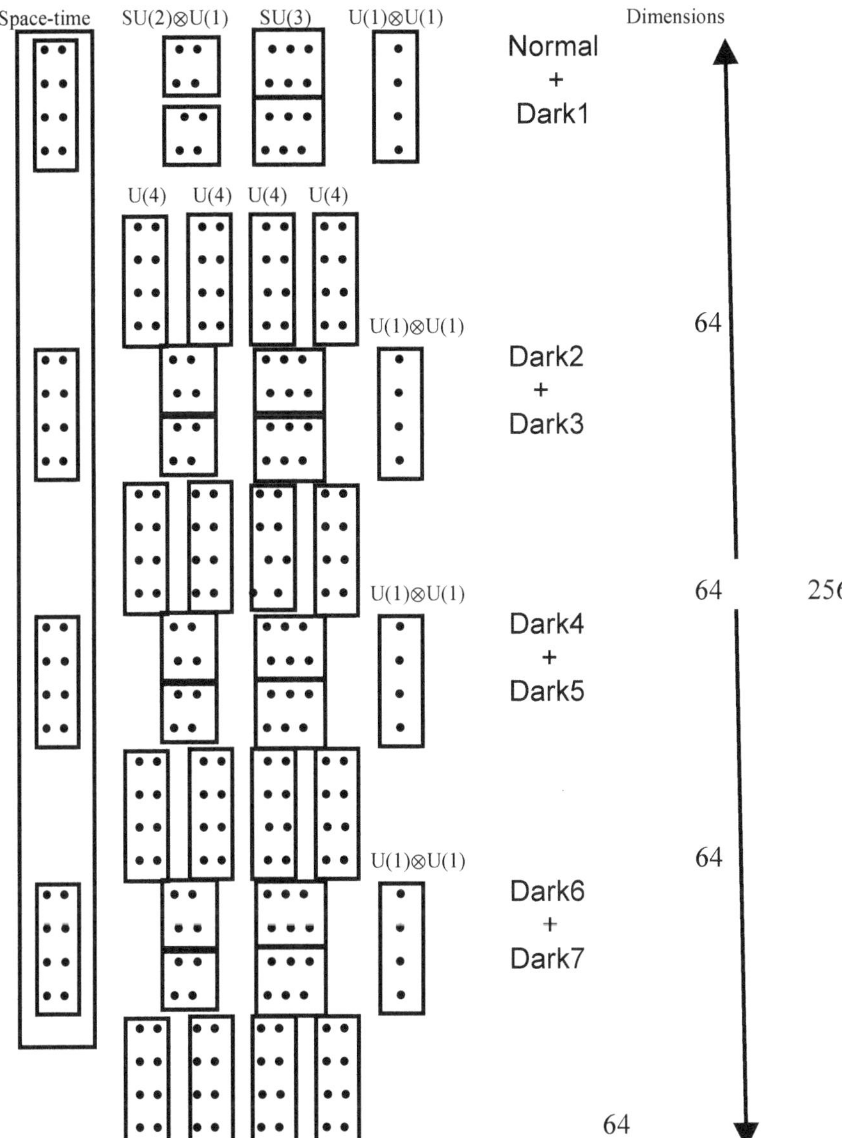

Figure 5.5 The 256 dimension groups of the four 8 × 8 blocks in each section of Fig. 5.4 Each 256 dimension section is one of the four layers of UTMOST dimensions.

6. The BMOST – UTMOST Connection

Universes are embedded within the Megaverse. A QUeST universe has four complex quaternion space-time dimensions. The UTMOST Megaverse has eight complex octonion space-time dimensions. The BMOST space has 10 space-time dimensions.

We can embed any number of QUeST universes within the UTMOST Megaverse.

We found we could define a basis for UTMOST, called BMOST, consisting of one seed urfermion residing in a ten dimension space-time (Chapter 4). a seed urfermion-antiurfermion pair annihilates into a Megaverse particle. This 10-space may be related to a yet deeper SuperString theory. Its details remain to be understood.

Thus a general foundation exists for EVERYTHING in ten dimension urfermions.

Appendix A. Enhanced Quantum Field Theory: Two Tier & PseudoQuantum from Chapters 40 and 41 of Blaha (2020c)

This Appendix describes Two-Tier Quantum Coordinates (Chapter 40 below) and PseudoQuantum field Theory (Chapter 41 below).

We begin by outlining the advantages of PseudoQuantum Field Theory (from Blaha (2020c) and earlier books). To these advantages we add the capability of supporting the derivation of QUeST from BQUeST and the derivation of UTMOST from BMOST. The advantages are:

1. Quantization in any coordinate system in flat or curved space-times with an invariant definition of asymptotic particle states. An n particle asymptotic state in one coordinate system is a unitarily equivalent n particle asymptotic state in any other coordinate system. Therefore particle number is invariant under change of coordinate system. This is important for the Unified SuperStandard Theory in curved space-times. It is also important for quantization in higher dimensional Euclidean spaces such as the Megaverse. The method was developed in the late 1970's by the author to provide a quantization procedure which supports a unique particle interpretation of states in arbitrary non-static space-times where no global timelike coordinate (Killing vector) exists. PseudoQuantum Field Theory which we developed in a series of books[49] also can be formulated in the Megaverse. Thus we can use it in the Megaverse to implement the Higgs Mechanism to generate particle masses and symmetry breaking.

2. PseudoQuantum Field Theory enables one to define Higgs particle dynamics in such a way that a non-zero vacuum expectation value cleanly separates from the quantum field part of the Higgs fields. This technique can be used in symmetry breaking mechanisms, mass generation, and possible generation of coupling constants as vacuum expectation values.

3. It supports the canonical definition of higher derivative field theories through the use of the Ostrogradski bootstrap. See Appendix B where a fourth order theory of the Strong interaction is defined that has color confinement and a linear r potential. The potential part of this theory was used by the Cornell group to calculate the Charmonium spectrum. (See Blaha (2017b) for details.)

---

[49] See Blaha (2017b) for the discussion of the PseudoQuantum field theory formalism for Higgs particles in our Extended Standard Model. See chapter 20 of Blaha (2017b), and earlier books, for a more detailed view than that presented here.

An associated advantage of using PseudoQuantum Field Theory is that it provides for retarded propagators and an Arrow of Time.
AND

4. Support for derivations of QUeST and UTMOST from one dimension BQUeST and BMOST.

See the author's paper, S. Blaha, Il Nuovo Cimento **49A**, 35 (1979). Other papers from the 1970's are listed therein. Blaha (2007b) and (2011c) describe PseudoQuantum Field theory.

Blaha (2016f) describes PseudoQuantum Mechanics (*CQ Mechanics*), a related formalism, and describes applications (including the Schrödinger equation, Fokker-Planck equation, Boltzmann equation, and the Vlasov equation) that smoothly provide *continuous* transitions between Quantum Theory and Classical Theory.

We now turn to describing Two-Tier Quantum Coordinates.

40. Two-Tier Coordinates

Originally Two-Tier coordinates were developed by this author to remove infinities that appear in perturbation theory calculations. We showed that the quantum smeared coordinates of Two-Tier Quantum Field Theory succeeded in removing all ultra-violet infinities in perturbation theory including the fermion triangle infinities. Remarkably the high precision, low energy[50] predictions of QED remained true in Two-Tier QED and thus remained consistent with experiment to a hitherto unsurpassed level of accuracy. 'Low' energy predictions in other quantum field theories also remained unchanged. At high energies, Two-Tier perturbation theory results are finite and consequently all ultra-violet infinities, to any order in perturbation theory, in *any number of space-time dimensions* were eliminated.

In addition to removing perturbation theory infinities Two-Tier coordinates enable us to define finite theories of Quantum Gravity and 'non-renormalizable' quantum field theories based on polynomial lagrangians, to tame vacuum fluctuations, to eliminate infinities associated with the Big Bang, and possibly to generate the explosive growth of the universe in its role as Dark Energy.[51]

Two-Tier Quantum Field Theory is established on the most fundamental level.

40.1 Two-Tier Features in 4-Dimensional Space-Time

Two-Tier Quantum Field Theory,[52] which was based on a new method[53] in the Calculus of Variations, uses two sets of fields to introduce quantum coordinates. We

[50] Relative to a mass scale that was perhaps of the order of the Planck mass.
[51] See Blaha (2017b) and earlier books for details. This section is basically a summary of some features.
[52] See Blaha (2005a), and Blaha (2002), for discussions of this new method to eliminate infinities in quantum field theory calculations.

shall consider this technique for the specific case of a massless vector field $V^i(y)$ analogous to the electromagnetic field.

In 4-dimensional space-time the massless vector field has the form $Y^\mu(y)$ where the index μ ranges from 0 through 3. The X^μ coordinate system, where it appears, has a c-number real part and a q-number imaginary part. Thus particle fields which are normally defined on four-dimensional real space-time will now be defined on a complex four-dimensional space-time where four imaginary dimensions will appear as *Quantum Dimensions* embodied in a vector quantum field $Y^\mu(y)$:

$$X^\mu(y) = y^\mu + i\, Y^\mu(y)/M_c^2$$

where M_c is an extremely large mass of the order of the Planck mass or perhaps much larger.

The $Y^\mu(y)$ field is a function of the subspace y coordinates. The real part of the space-time dimensions will be taken to be the space of real-valued y coordinates.[54]

The imaginary part of space-time coordinates is the a massless $Y^\mu(y)$ vector quantum field that is suppressed further by a very large mass scale – perhaps of the order of the Planck mass – that reduces the imaginary Quantum Dimensions to the infinitesimal except at large momenta. The effects of Quantum Dimensions only become appreciable in quantum field theory at energies of the order of M_c. At these energies exponential Gaussian factors in each particle (and ghost) propagator are generated by the Quantum Dimensions and serve to make perturbation theory calculations ultra-violet finite – including calculations in Quantum Gravity.

The formalism introduces a new form of interaction that does not have the form of the simple polynomial interactions that have hitherto dominated quantum field theories. This form of interaction takes place via the composition of quantum fields and can be called a *Dimensional Interaction,* or an *Interdimensional Interaction,* since it affects particle behavior through Quantum Dimensions.

The basic ansatz of the Two-Tier formalism is to replace every appearance of a coordinate x in a quantum field with the variable

$$x^\mu \rightarrow X^\mu = (y^0, \mathbf{y} + \mathbf{Y}(y^0, \mathbf{y})/M_c^2)$$

where $\mathbf{Y}(y^0, \mathbf{y})$ is the spatial part of a free massless vector field with features that are identical to the free QED field in the Radiation gauge.

Then one finds that the momentum space free field Feynman propagators G(k) of all particles acquires a Gaussian factor exp(h(k)):

$$G(k) \rightarrow G(k)\, \exp(h(k))$$

---

[53] Blaha (2005a) describes our method for the composition of extrema in some detail.

[54] In a deeper theory the real part might also be a quantum field that undergoes a condensation to generate c-number coordinates. We will not consider this possibility in this book.

so that all perturbation theory diagrams are finite. The result is finite perturbative results for all calculations to any order in perturbation theory. Blaha (2005a) shows that Two-Tier theories are finite, Poincare covariant, and unitary. (See Blaha (2005a), chapter 5, for a complete discussion.)

40.2 Simple Two-Tier X^μ Formalism

In this subsection we will describe the basic Two-Tier formalism. Taking the lagrangian described in Blaha (2005a):[55]

$$\mathscr{L}(y) = \mathscr{L}_F(X^\mu(y))J + \mathscr{L}_C(X^\mu(y), \partial X^\mu(y)/\partial y^\nu, y) \tag{40.1}$$

where

$$X^\mu(y) = y^\mu + i\, Y^\mu(y)/M_c^2 \tag{40.2}$$

with M_c being a large mass scale, $Y_\mu(y)$ a vector quantum field, and where J is the absolute value of the Jacobian of the transformation from X to y coordinates:

$$J = |\partial(X)/\partial(y)|$$

The lagrangian term $\mathscr{L}_C$ is

$$\mathscr{L}_C = +\tfrac{1}{4} M_c^4 F^{\mu\nu}F_{\mu\nu}$$

with

$$\begin{aligned} F_{\mu\nu} &= \partial X_\mu/\partial y^\nu - \partial X_\nu/\partial y^\mu \\ &\equiv i\,(\partial Y_\mu/\partial y^\nu - \partial Y_\nu/\partial y^\mu)/M_c^2 \end{aligned} \tag{40.3}$$

The lagrangian term $\mathscr{L}_F(X^\mu(y))$ contains the terms for scalar, fermion and other gauge terms in general. The sign in $\mathscr{L}_C$ is not negative – contrary to the conventional electromagnetic Lagrangian. The reason for this difference is that the quantum field part of X^μ is imaginary. Thus $\mathscr{L}_C$ ends up having the correct sign after taking account of the factor of i in the field strength $F_{\mu\nu}$.

Defining

$$F_{Y_{\mu\nu}} = (\partial Y_\mu/\partial y^\nu - \partial Y_\nu/\partial y^\mu)$$

we see the Lagrangian assumes the form of the conventional electromagnetic Lagrangian:

$$\mathscr{L}_C = -\tfrac{1}{4} F_Y^{\mu\nu}F_{Y_{\mu\nu}}$$

The action of this theory has the form

$$I = \int d^4y\, \mathscr{L}(y)$$

---

[55] Eq. 7.1.

40.3 Y^μ Gauge

The gauge invariance of the Lagrangian allows us to choose a convenient gauge. The gauge invariance of the full Lagrangian

$$\mathscr{L}_S = L_F(\phi(X), \partial\phi/\partial X^\mu)\, J + \mathscr{L}_C(X^\mu(y), \partial X^\mu(y)/\partial y^\nu)$$

is based on the standard gauge invariance of $\mathscr{L}_C$, and the gauge invariance of $J\mathscr{L}_F$ in the form of translational invariance

$$X^\mu(y) \rightarrow X^\mu(y) + \delta X^\mu(y)$$

for the special case of a translation of X with the form of a gauge transformation:

$$\delta X^\mu(y) = \partial\Lambda(y)/\partial y_\mu$$

In this case we find

$$\int d^4y\ \Lambda(y)\, \partial\, [\, J\, \partial/\partial X^\mu\, \mathscr{T}_{F\mu\nu}\,]/\partial y_\nu\ = 0 \tag{40.4}$$

after a partial integration and so we have the differential conservation law:

$$\partial\, [\, J\, \partial \mathscr{T}_{F\mu\nu}/\partial X^\mu]/\partial y_\nu\ = 0$$

since $\Lambda(y)$ is arbitrary. This conservation law is trivially obeyed:

$$\partial \mathscr{T}_{F\mu\nu}/\partial X^\mu = 0 \tag{40.5}$$

Thus translational invariance in the $\mathscr{L}_F$ sector together with standard gauge invariance in the $\mathscr{L}_C$ sector automatically guarantees Y field gauge invariance of the total Lagrangian. We use the separate invariance of each term of

$$L = \int d^4y\, [\mathscr{L}_F\, J + \mathscr{L}_C\,] = \int d^4X\, \mathscr{L}_F + \int d^4y\, \mathscr{L}_C = L_F + L_C$$

under a constant translation $X^\mu \rightarrow X^\mu + \delta X^\mu$ where δX^μ is constant. Then we consider a position dependent translation/gauge transformation, which taken together with the above equation, establishes the invariance under the position dependent translation/gauge transformation.

An alternate approach that leads to the same result is to start with the particle part of the Lagrangian $\mathscr{L}_F$ rewritten to be invariant under general coordinate transformations, as it must, when we generalize to include General Relativity. Since position dependent translations are a form of general coordinate transformation the full theory must be invariant under position dependent translations due to invariance under general coordinate transformations.

Having established invariance under gauge transformations we now choose to use the most convenient gauge – the radiation gauge[56]:

$$\partial Y^i/\partial y^i = 0 \tag{40.6}$$

where i = 1, 2, 3, which, in the absence of external sources, allows us to set

$$Y^0 = 0$$

since Y^0 does not have a canonically conjugate momentum. A conventional treatment leads to the equal time commutation relations:

$$[Y^\mu(\mathbf{y}, y^0), Y^\nu(\mathbf{y}', y^0)] = [\pi^\mu(\mathbf{y}, y^0), \pi^\nu(\mathbf{y}', y^0)] = 0 \tag{40.7}$$

$$[\pi^j(\mathbf{y}, y^0), Y_k(\mathbf{y}', y^0)] = -i\, \delta^{tr}{}_{jk}(\mathbf{y} - \mathbf{y}')$$

(Note the locations of the j indexes above introduce a minus sign.) where

$$\pi^k = \partial \mathscr{L}_C/\partial Y_k'$$
$$\pi^0 = 0$$

$$\delta^{tr}{}_{jk}(\mathbf{y} - \mathbf{y}') = \int d^3k\, e^{i\,\mathbf{k}\cdot(\mathbf{y} - \mathbf{y}')}(\delta_{jk} - k_j k_k/\mathbf{k}^2)/(2\pi)^3$$

$$Y_k' = \partial Y_k/\partial y^0$$

The Radiation gauge reveals the two degrees of freedom that are present in the vector potential. The Fourier expansion of the vector potential is:

$$Y^i(y) = \int d^3k\, N_0(k) \sum_{\lambda=1}^{2} \varepsilon^i(k, \lambda)[a(k,\lambda)\, e^{-ik\cdot y} + a^\dagger(k,\lambda)\, e^{ik\cdot y}] \tag{40.8}$$

where

$$N_0(k) = [(2\pi)^3 2\omega_k]^{-1/2}$$

and (since m = 0)

$$\omega_k = (\mathbf{k}^2)^{1/2} = k^0$$

[56] It is also possible to quantize using an indefinite metric that preserves manifest Lorentz covariance as was done by Gupta and Bleuler for the electromagnetic field. We will use the Gupta-Bleuler approach later to establish covariance under special relativity later. Now we opt for manifest positivity and use the radiation gauge.

with $\vec{\epsilon}(k, \lambda)$ being the polarization unit vectors for $\lambda = 1,2$ and $k^\mu k_\mu = 0$.

The further development of this theory is described in Part 3 of Blaha (2005a).

40.4 Scalar Field Quantization Using X^μ

We will begin by considering the case of a scalar quantum field theory. We assume a real underlying y subspace. Since X^μ is a set of coordinates, we choose to define a scalar field ϕ as a function of X^μ, which, in turn, is a function of the y^ν coordinates. We will provisionally second quantize ϕ treating X^μ as c-number coordinates using a conventional approach.[57]

We assume a Lagrangian, with the momentum conjugate to ϕ:

$$\pi_\phi = \partial L_F / \partial\phi' \equiv \partial L_F / \partial(\partial\phi/\partial X^0) \tag{40.9}$$

Following the canonical quantization procedure, π and ϕ become hermitian operators with equal time ($X^0 = X^{0\prime}$) commutation rules:

$$\begin{aligned}[]
[\phi(X), \phi(X')] &= [\pi_\phi(X), \pi_\phi(X')] = 0 \\
[\pi_\phi(X), \phi(X')] &= -i\,\delta^3(\mathbf{X} - \mathbf{X}')
\end{aligned} \tag{40.10}$$

The standard Fourier expansion of the solution to the Klein-Gordon equation is:

$$\phi(X) = \int d^3p\, N_m(p)\, [a(p)\, e^{-ip\cdot X} + a^\dagger(p)\, e^{ip\cdot X}]$$

where

$$N_m(p) = [(2\pi)^3 2\omega_p]^{-1/2}$$

and

$$\omega_p = (\mathbf{p}^2 + m^2)^{1/2}$$

The commutation relations of the Fourier coefficient operators are:

$$\begin{aligned}[]
[a(p), a^\dagger(p')] &= \delta^3(\mathbf{p} - \mathbf{p}') \\
[a^\dagger(p), a^\dagger(p')] &= [a(p), a(p')] = 0
\end{aligned}$$

The reader will recognize the quantization procedure is formally identical to the standard canonical quantization procedure of a free scalar quantum field.

---

[57] Some texts are: Bogoliubov, N. N., Shirkov, D. V., *Introduction to the Theory of Quantized Fields* (Wiley-Interscience Publishers Inc., New York, 1959); Bjorken, J. D., Drell, S. D., *Relativistic Quantum Fields* (McGraw-Hill, New York, 1965); Huang, K., *Quarks, Leptons & Gauge Fields Second Edition* (World Scientific, River Edge, NJ, 1992); Kaku, M., *Quantum Field Theory* (Oxford University Press, New York, 1993); Weinberg, S., *The Quantum Theory of Fields* (Cambridge University Press, New York, 1995).

In the case of spin ½, spin 1 and spin 2 fields the standard quantization procedure *in terms of the X coordinate system* can also be followed in a way similar to the procedure in standard texts.

40.5 Scalar Feynman Propagators

The momentum space free field Feynman propagators G…(k) of all particles and ghosts in all Two-Tier Quantum Field Theories acquires a Gaussian factor exp(h(k)):

$$G\ldots(k) \rightarrow G\ldots(k)\exp(h(k))$$

so that all perturbation theory diagrams are finite. The result is a finite perturbative result in all calculations to any order in perturbation theory. Blaha (2005a) shows that Two-Tier theories are finite, Poincare covariant, and unitary.

An example of the Two-Tier effect on propagators is the case of the Two-Tier photon propagator. The Two-Tier photon propagator[58] is:

$$iD_F^{TT}(y_1 - y_2)_{\mu\nu} = -i\int \frac{d^4p\, e^{-ip\cdot z}\, g_{\mu\nu} R(\mathbf{p}, z)}{(2\pi)^4 (p^2 + i\varepsilon)} \tag{40.11}$$

(since the imaginary parts can be taken to be zero: $y_{1i}^{\mu} - y_{2i}^{\mu} = 0$) where

$$z^{\mu} = y_{1r}^{\mu} - y_{2r}^{\mu}$$

$$R(\mathbf{p}, \mathbf{z}) = \exp[-p^i p^j \Delta_{Tij}(z)/M_c^4]$$

$$= \exp\{ -\mathbf{p}^2[A(v) + B(v)\cos^2\theta] / [4\pi^2 M_c^4|\mathbf{z}|^2$$

with i, j = 1, 2, 3, and with $\Delta_{Tij}(z)$ being the commutator of the positive frequency part $Y^+_k(y)$ and the negative frequency part $Y^-_k(y)$ of $Y_k(y)$:

$$\Delta_{Tij}(z) = [Y^+_j(y_{1r}), Y^-_k(y_{2r})] = \int d^3k\, e^{ik\cdot(y_{1r} - y_{2r})} (\delta_{jk} - k_j k_k/\mathbf{k}^2)/[(2\pi)^3 2\omega_k] \tag{40.12}$$

and

$$v = |z^0|/|\mathbf{z}|$$
$$A(v) = (1 - v^2)^{-1} + .5v \ln[(v - 1)/(v + 1)]$$
$$B(v) = v^2(1 - v^2)^{-1} - 1.5v \ln[(v - 1)/(v + 1)]$$
$$\mathbf{p}\cdot\mathbf{z} = |\mathbf{p}|\, |\mathbf{z}| \cos\theta$$

---

[58] Blaha (2005a).

with $|\mathbf{p}|$ denoting the length of a spatial vector **p**, $|\mathbf{z}|$ denoting the length of a spatial vector **z**, and with $|z^0|$ being the absolute value of z^0.

The gaussian factors R(**p**, z) which appear in all Two-Tier propagators damp the large momentum behavior of all perturbation theory integrals producing a completely finite perturbation theory and yet give the usual results of perturbation theory at energies that are small compared to the mass scale M_c.

40.6 String-like Substructure of the Theory

Two-tier Quantum field Theory endows each particle with an extended structure that resembles the extended structure seen in boson string and Superstring theories. For example, Bailin (1994) use the operator[59]

$$V_\Lambda(k) = \int d^2\sigma \sqrt{-h}\, W_\Lambda(\tau, \sigma)\, e^{-ik\cdot X}$$

where X^μ is a quantized fourier expansion of the string fields (see eq. 7.22 of Bailin (1994)).

We note our X^μ coordinate-field has two transverse degrees of freedom due to gauge invariance, which also invites comparison to the boson string. A point of difference is that we have a well-defined quantum field theoretic formulation in conventional space-time that has the Standard Model as its "large distance" behavior thus introducing a note of reality that is not apparent in Superstring theories. We see that the interacting quantum field theories based on this approach also have good, finite, short distance behavior just as string theories.

The scalar, and other particles', Feynman propagators can be viewed as describing the propagation of a particle cloaked (accompanied) by a cloud of Y particles (which generates the R(**p**, $y_1 - y_2$) factor in the above propagator). If we examine the fourier transform of R(p, z) we see:

$$(2\pi)^4 R(\mathbf{p}, q) = \int d^4z\, e^{iq\cdot z}\, R(\mathbf{p}, z) = \int d^4z\, e^{iq\cdot z} \exp[\, -p^i p^j \Delta_{Tij}(z)/M_c^{\,4}] \qquad (40.13)$$

and we find

$$R(\mathbf{p},q) = \sum_{n=0}^{\infty} [i(2\pi M_c)^4]^{-n} (n!)^{-1} \prod_{j=1}^{n} [\int d^4k_j\, \theta(k_j^{\,0})(\mathbf{p}^2 - (\mathbf{p}\cdot\mathbf{k}_j)^2/\mathbf{k}_j^{\,2})/(k_j^{\,2} + i\varepsilon)]\, \delta^4(q - \sum k_r)$$

which can be interpreted as a "cloud" of Y particles dressing the "bare" particle propagator. (The apparent divergences for R(p, q) are an artifact of the expansion and the subsequent fourier transformation. They are not present in the R(**p**, $y_1 - y_2$) factor in the propagator. See Fig. 40.1 for the Feynman diagram of the Two-Tier 'cloaked' propagator as compared to the normal scalar particle Feynman propagator. The Two-

[59] D. Bailin and A. Love, *Supersymmetric Gauge Field Theory and String Theory* (Institute of Physics Publishing, Philadelphia, PA, 1994) page 272.

Tier Feynman propagator is basically a conventional scalar propagator that is modified by coherent Y particle emission.[60]

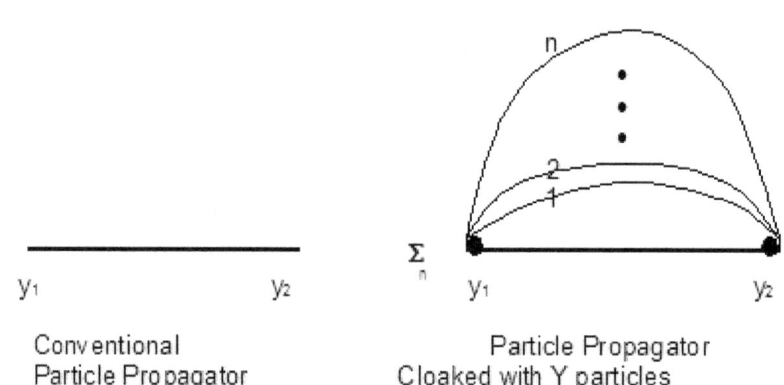

Figure 40.1. Feynman diagram for conventional and the nth diagram of a cloaked Two-Tier propagator.

We note that R(p, q) satisfies the convolution theorem:

$$\int d^4k\, R(\mathbf{p}, k)\, R(\mathbf{p}, q-k) = [R(\mathbf{p}, q)]^2$$

or

$$(2\pi)^4 \int d^4z\, e^{iq\cdot z}\, R(\mathbf{p}, z)\, R(\mathbf{p}, z) = [\int d^4z\, e^{iq\cdot z}\, R(\mathbf{p}, z)\,]^2 \qquad (40.14)$$

The proof follows from the Binomial theorem.

40.7 Two-Tier Complexon Quantum Fields

In the case of the Complexon Standard Model we will need two variables X_r^μ and X_i^μ since we have complex spatial 3-coordinates. We define them similarly to the previous case:

$$X_r^\mu(y_r) = y_r^\mu + i\, Y_r^\mu(y_r)/M_c^2$$
$$X_i^\mu(y_i) = y_i^\mu + i\, Y_i^\mu(y_i)/M_c^2$$

where we choose the same mass scale for both the “real” and “imaginary” variables. The Two-Tier, single generation, version of the Complexon Standard Model then has an action of the form

$$I_{CSMtt} = \int dy^0 d^3y_r d^3y_i\, (\mathscr{L}_{CSM}(X_r^\mu(y_r), \mathbf{X}_i^k(y_i))J_2)|_{yi0 = 0,\ Yr0 = Yi0 = 0} + \qquad (40.15)$$
$$+ \int dy_r^0 d^3y_r\, \mathscr{L}_C(X_r^\mu(y_r), \partial X_r^\mu(y_r)/\partial y_r^\nu, y_r) +$$

[60] T. W. B. Kibble, Phys. Rev. **173**, 1527 (1968) and references therein. In particular see p. 1532 of Kibble’s paper.

$$+ \int dy_i^0 d^3y_i \mathscr{L}_C(X_i^\mu(y_i), \partial X_i^\mu(y_i)/\partial y_i^\nu, y_i)$$

where the replacements

$$x^\mu \equiv x_r^\mu \rightarrow X_r^\mu(y_r)$$
$$x_i^k \rightarrow X_i^k(y_i)$$

for μ = 0, 1, 2, 3 and k = 1, 2, 3 are made, followed by defining $y_r^0 = y^0$ and making a Complex Lorentz transformation to a frame where $y_i^0 = 0$. J_2 is the absolute value of the Jacobian of the transformation from (X_r, X_i) to (y_r, y_i) coordinates:

$$J_2 = |\partial(X_r, X_i)/\partial(y_r, y_i)|$$

We also choose gauges where $Y_r^0 = Y_i^0 = 0$. These types of transformations and gauge choices are discussed in detail in Blaha (2005a). The lagrangian terms $\mathscr{L}_C(X_r^\mu(y_r), \partial X_r^\mu(y_r)/\partial y_r^\nu, y_r)$ and $\mathscr{L}_C(X_i^\mu(y_i), \partial X_i^\mu(y_i)/\partial y_i^\nu, y_i)$ have the same form:

$$\mathscr{L}_C = +\tfrac{1}{4} M_c^4 F^{\mu\nu} F_{\mu\nu} \qquad (40.16)$$

with

$$F_{\mu\nu} = \partial X_\mu/\partial y^\nu - \partial X_\nu/\partial y^\mu$$
$$\equiv i\,(\partial Y_\mu/\partial y^\nu - \partial Y_\nu/\partial y^\mu)/M_c^2$$

or defining

$$F_{Y_{\mu\nu}} = (\partial Y_\mu/\partial y^\nu - \partial Y_\nu/\partial y^\mu)$$

we see each lagrangian assumes the form of the conventional electromagnetic Lagrangian:

$$\mathscr{L}_C = -\tfrac{1}{4} F_Y^{\mu\nu} F_{Y_{\mu\nu}}$$

The lagrangian is supplemented with the following condition on all complexon fields $\Phi_{\ldots}$:

$$(\partial/\partial X_r^k(y_r))\,(\partial/\partial X_i^k(y_i))\Phi\ldots = 0 \qquad (40.17)$$

summed over k = 1, 2, 3. Non-complexon fields Ω… in our left-handed formulation satisfy the subsidiary condition:

$$\{(\partial/\partial X_r^k(y_r))(\partial/\partial X_i^k(y_i)) - [(\partial/\partial X_r^k(y_r))^2(\partial/\partial X_i^m(y_i))^2]^{1/2}\}\Omega\ldots = 0 \qquad (40.18)$$

summed over k = 1, 2, 3 and over m = 1, 2, 3 separately in each of the two terms.

40.8 Complexon Feynman Propagator

In the case of complexons, the Two-Tier Feynman propagator differs from the non-complexon case by having an integration over imaginary spatial 3-momenta, a derivative of a delta function embodying the orthogonality of the real and imaginary 3-momenta, and two factors of R(**p**, z): one factor being R($\mathbf{p}_r$, z_r) and the other factor being R($\mathbf{p}_i$, z_i) (where the time components $z_r^0 = z^0$ and $z_i^0 = 0$ since there is only one real time coordinate[61]) thus providing large momentum convergence for both real and imaginary 3-momentum integrations.

For a normal scalar particle the Feynman propagator is:

$$i\Delta_{CTF}(x-y) = \theta(x^+ - y^+)\langle 0|\phi_{CT}(x)\,\phi_{CT}(y)|0\rangle + \theta(y^+ - x^+)\langle 0|\phi_{CT}(y)\phi_{CT}(x)|0\rangle$$
$$= i\int d^4p_r d^3p_i (2\pi)^{-7}\delta'(\mathbf{p}_r\cdot\mathbf{p}_i/m^2)e^{-ip^+(x^- - y^-) - ip^-(x^+ - y^+) + i\mathbf{p}_\perp\cdot(\mathbf{x}_\perp - \mathbf{y}_\perp) - i\mathbf{p}_i\cdot(\mathbf{x}_i - \mathbf{y}_i)}/(p^2 + m^2 + i\varepsilon) \qquad (40.19)$$

in conventional quantum field theory.

In the case of Two-Tier quantum field a scalar *complexon* particle has the Feynman propagator

$$i\Delta_{CTFtt}(x-y) = i\int d^4p_r d^3p_i (2\pi)^{-7}\delta'(\mathbf{p}_r\cdot\mathbf{p}_i/m^2)\, R(\mathbf{p}_r, z_r)R(\mathbf{p}_i, z_i)\cdot \qquad (40.20)$$
$$\cdot e^{-ip^+(x^- - y^-) - ip^-(x^+ - y^+) + i\mathbf{p}_\perp\cdot(\mathbf{x}_\perp - \mathbf{y}_\perp) - i\mathbf{p}_i\cdot(\mathbf{x}_i - \mathbf{y}_i)}/(p^2 - m^2 + i\varepsilon)$$

where the time components $z_r^0 = z^0$ and $z_i^0 = 0$ since there is only one time coordinate, where R(**p**, z) is given in the previous subsection, and where $p^2. = p^{0\,2} - p_r^2 + p_i^2$.

Propagators for other types of particles are similarly modified in the Two-Tier formalism (See Blaha 2005a).

40.9 Vacuum Fluctuations

While the expectation value of a *conventional* free scalar field $\phi_{conv}(x)$ is zero in a conventional quantum field theory:

$$\langle 0|\phi_{conv}(x)|0\rangle = 0 \qquad (40.21)$$

the vacuum fluctuations of *conventional* scalar quantum field theory are quadratically divergent:

$$\langle 0|\phi_{conv}(x)\phi_{conv}(x)|0\rangle = \int d^3p/[(2\pi)^3 2\omega_p] \qquad (40.22)$$

In "Two-Tier" quantum field theory we find the vacuum expectation value of a free field is zero *and the expectation value of the square of the field is also zero:*

---

[61] We can arrange for $z_i^0 = 0$ by making a Complex Lorentz transformation to an inertial frame where z is real.

$$<0|\phi(X)\phi(X)|0> = \int d^3p\, e^{-p^ip^j\Delta_{Tij}(0)/Mc^4}/[(2\pi)^3 2\omega_p] = 0$$

since the exponential factor in the integral is $-\infty$. The exponent contains

$$\Delta_{Tij}(z) = \int d^3k\, e^{-ik.z}\,(\delta_{ij} - k_ik_j/\mathbf{k}^2)/[(2\pi)^3 2\omega_k] \qquad (40.23)$$

where "T" is for "Two-Tier". Thus *vacuum fluctuations are zero in Two-Tier quantum field theory*. Correspondingly, we will see that renormalization constants are finite in the Two-Tier versions of QED, Electroweak Theory, the Standard Model and Quantum Gravity. See Blaha (2017b) and references therein for more details.

40.10 Time Intervals in General Relativity

Wigner[62] has studied the measurement of time intervals in General Relativity and sees a problem in the measurement of extremely short intervals. According to Wigner, the measurement of a time interval in a region of space requires the measurement of the length of time required for an event to happen. The measurement requires an accurate clock. But the accuracy of the clock is limited by the energy-time uncertainty relation:

$$\Delta E \Delta t \geq \hbar \qquad (40.24)$$

Thus the uncertainty in the clock's time measurement is related to the uncertainty in the clock's energy which is, in turn, related to the uncertainty in the clock's mass:

$$\Delta E = (\Delta m)c^2$$

To obtain "infinite" accuracy the uncertainty (fluctuations) in the clock's mass must be infinite and thus the clock's mass must be infinite. Infinite fluctuations in the clock's mass will produce corresponding infinite fluctuations in the gravitational field.

$$\Delta h \propto \Delta E \qquad \text{(in conventional General Relativity)}$$

As a result the notions of space-time and time intervals (which depend on the geometry through General Relativity) become uncertain. Thus, according to Wigner, and others, the concept of time intervals and space-time points becomes questionable.

The Two-Tier version of Quantum Gravity offers a way out of this dilemma. The gravitational force becomes stronger as one goes to shorter distances (higher energies) down to a distance (or up to an energy) whose scale is set by M_c. At shorter distances (higher energies) the gravitational force becomes weaker and declines to zero at zero distance. Thus at very high energy the gravitational field fluctuations (Δh) are at worst inversely proportional to the energy (and probably decline by a higher power of inverse energy.) (The same considerations would apply if one chooses to consider fluctuations in the Riemann-Christoffel symbols.)

---

[62] E. P. Wigner, Rev. Mod. Phys. **29**, 255 (1957); J. Math. Phys. **2**, 207 (1961).

$$\Delta h < c_1/E < c_1/(\Delta E) \quad \text{(in Two-Tier Quantum Gravity)} \qquad (40.25)$$

where c_1 is a constant. Thus Wigner's conclusion does not hold in the Two-Tier version of Quantum Gravity as gravitational fluctuations actually become smaller at energies above a critical energy whose scale is set by M_c.

In fact, combining the above equations we see

$$c_1\Delta t/\Delta h \geq \hbar$$

at sufficiently high energy. Therefore the time uncertainty Δt, and the gravitational field fluctuations Δh, can both decrease while maintaining the energy-time uncertainty relation. *Thus the notion of a space-time point "is saved" in Two-Tier quantum gravity.*

40.11 Vacuum Fluctuations in the Gravitation Fields

While the expectation value of the free graviton field $h_{\mu\nu\text{conv}}(x)$ (weak field approximation) is zero in a conventional quantum field theoretic approach:

$$<0|h_{\mu\nu\text{conv}}(x)|0> = 0 \qquad (40.26)$$

the vacuum fluctuations of the *conventional* quantum graviton field is quadratically divergent since

$$<0|h_{\mu\nu\text{conv}}(x)h_{\alpha\beta\text{conv}}(x)|0> = \int d^3p\, b'_{\mu\nu\alpha\beta}(p)/[(2\pi)^3\, 2\omega_p] = \infty \qquad (40.27)$$

where $b'_{\mu\nu\alpha\beta}(p)$ is a rational function of the momentum p.

In "Two-Tier" quantum field theory we find

$$<0|h_{\mu\nu}(X)h_{\alpha\beta}(X)|0> = \int d^3p\, b'_{\mu\nu\alpha\beta}(p)\, e^{-p^ip^j\Delta_{Tij}(0)}/[(2\pi)^3 2\omega_p] = 0 \qquad (40.28)$$

since the exponential factor in the integrand is $-\infty$. The exponent contains

$$\Delta_{Tij}(z) = \int d^3k\, e^{-ik\cdot z}(\delta_{ij} - k_ik_j/\mathbf{k}^2)/[(2\pi)^3 2\omega_k]$$

Thus the vacuum fluctuations of $h_{\mu\nu}$ are zero in "Two-Tier" quantum field theory and, correspondingly, the weak field Two-Tier quantization of Quantum Gravity is consistently finite (and weak in perturbation theory calculations.)

40.12 Two-Tier Features in D-Dimensional Space-Time (such as the Megaverse)

Since a field, quantized in D-dimensional conventional coordinates (D > 4), would lead to divergences in perturbation theory calculations, we can use D-dimensional Two-Tier coordinates to avoid divergences in perturbation theory:

$$Y^i(y) = y^i + i\, Y_u^{\,i}(y)/M_u^{\,D/2} \qquad (40.29)$$

where $Y_u^{\,i}(y)$ for i = 1, …, D is a D-dimensional free gauge field and M_u is a mass of the order of the Planck mass or greater. The $Y_u^{\,i}(y)$ term adds a quantum field to the D coordinates making them a set of quantum coordinates. Quantum coordinate derivatives are defined by

$$\partial_i = \partial/\partial Y^i(y) = \partial/\partial(y^i - Y_u^{\,i}(y)/M_u^{\,D/2}) \qquad (40.30)$$

The use of these coordinates to quantize particle fields leads to a completely finite perturbation theory. We applied them in Blaha (2017b) to create a finite fundamental theory of mater. We applied them to fields in the Megaverse[63] to achieve a finite theory of Megaverse dynamics for elementary particles and universe particles.

The second quantization of a vector gauge field $V^i(y)$ is analogous to the second quantization of the electromagnetic field. The lagrangian density terms for the free $V^i(Y(y))$ fields is

$$\mathscr{L}_{Vu} = -\tfrac{1}{4} F_{Vu}^{\;ij}(Y(y))F_{Vuij}(Y(y)) \qquad (40.31)$$

The lagrangian is

$$L_{Vu} = \int d^Dy\, \mathscr{L}_{Vu}(Y(y))$$

with

$$F_{Vuij} = \partial V_i(Y(y))/\partial Y^j(y) - \partial V_j(Y(y))/\partial Y^i(y)$$

where the values of i and j range from 1 to D in this section.

The equal time commutation relations, using the Dth coordinate as the time coordinate, are specified in the usual way:

$$[V^i(Y(\mathbf{y}, y^0)), V^j(Y(\mathbf{y}', y^0))] = [\pi^i(Y(\mathbf{y}, y^0)), \pi^j(Y(\mathbf{y}', y^0))] = 0$$
$$[\pi_j(Y(\mathbf{y}, y^0)), V_k(Y(\mathbf{y}', y^0))] = -i\, \delta^{(D-1)tr}{}_{jk}(Y(\mathbf{y},0) - Y(\mathbf{y}',0))$$

where

$$\pi_u^{\,k} = \partial\mathscr{L}_{Vu}(V(Y(y)))/\partial V_k'(Y(y))$$
$$\pi_u^{\,D} = 0$$

for k = 1, … , (D – 1), and

$$\delta^{(D-1)tr}{}_{jk}(\mathbf{y} - \mathbf{y}') = \int d^{(D-1)}k\, e^{i\,\mathbf{k}\cdot(Y(\mathbf{y},0) - Y(\mathbf{y}',0))}\,(\delta_{jk} - k_jk_k/\mathbf{k}^2)/(2\pi)^{D-1} \qquad (40.32)$$
$$V_k'(Y(y)) = \partial V_k(Y(y))/\partial y^{1D}$$

for j, k = 1, 2, … , (D – 1).

If we choose the Radiation gauge for $V_k(Y(y))$:

---

[63] Blaha (2017c).

$$V^D(Y(y)) = 0$$
$$\partial V^j(Y(y))/\partial Y^j(y) = 0 \quad (40.33)$$

for j = 1, 2, … , (D – 1) then (D – 2) degrees of freedom (polarizations) are present in the vector potential.[64] The Fourier expansion of the vector potential $V^i(Y(y))$ is:

$$V^i(Y(y)) = \int d^{(D-1)}k\, N_{0V}(k) \sum_{\lambda=1}^{D-2} \varepsilon^i(k, \lambda)[a_V(k,\lambda) :e^{-ik.Y(y)}: + a_V^\dagger(k,\lambda) :e^{ik.Y(y)}:] \quad (40.34)$$

for i = 1, … , (D – 2) where

$$N_{0V}(k) = [(2\pi)^{(D-1)} 2\omega_k]^{-\frac{1}{2}}$$

and (since the field is massless)

$$k^D = \omega_k = (\mathbf{k}^2)^{\frac{1}{2}}$$

where k^D is the energy, and where the $\varepsilon^i(k, \lambda)$ are the polarization unit vectors for $\lambda = 1, \ldots, (D-2)$ and $k^\mu k_\mu = k^{D\,2} - \mathbf{k}^2 = 0$.

The commutation relations of the Fourier coefficient operators are:

$$[a_V(k,\lambda), a_V^\dagger(k',\lambda')] = \delta_{\lambda\lambda'}\delta^{D-1}(\mathbf{k} - \mathbf{k'})$$
$$[a_V^\dagger(k,\lambda), a_V^\dagger(k',\lambda')] = [a_V(k,\lambda), a_V(k',\lambda')] = 0$$

and the polarization vectors satisfy

$$\sum_{\lambda=1}^{D-2} \varepsilon_i(k, \lambda)\varepsilon_j(k, \lambda) = (\delta_{ij} - k_i k_j/\mathbf{k}^2)$$

The V^μ Feynman propagator is

$$iD_F^{trTT}(y_1 - y_2)_{jk} = \langle 0|T(V_j(Y(y_1))V_k(Y(y_2)))|0\rangle \quad (40.35)$$

$$= -ig_{jk} \int \frac{d^Dk\, e^{-ik.(y_1 - y_2)}\, R(\mathbf{k}, y_1 - y_2)}{(2\pi)^D\,(k^2 + i\varepsilon)}$$

where g_{jk} is the D-dimensional Lorentz metric and where $R(\mathbf{k}, y_1 - y_2)$ is given by

$$R(\mathbf{k}, y_1 - y_2) = \exp[-k^ik^j\Delta_{Tij}(y_1 - y_2)/M_u^D]$$
$$= \exp\{-k^2[A(v) + B(v)\cos^2\theta] / [(2\pi)^{D-2}M_u^4z^2]\}$$

where k^2 is *the sum of the squares of the D – 1 spatial components* with

---

[64] Note we use the Radiation gauge for $Y^\mu(y)$ also.

$$z^\mu = y_1{}^\mu - y_2{}^\mu$$
$$z = |\mathbf{z}| = |\mathbf{y}_1 - \mathbf{y}_2|$$
$$k = |\mathbf{k}|$$
$$v = |z^0|/z$$
$$A(v) = (1 - v^2)^{-1} + .5v \ln[(v - 1)/(v + 1)]$$
$$B(v) = v^2(1 - v^2)^{-1} - 1.5v \ln[(v - 1)/(v + 1)]$$
$$\mathbf{k}\cdot\mathbf{z} = kz \cos\theta$$

and $|\mathbf{k}|$ denoting the length of a spatial (D – 1)-vector **k** while $|z^0|$ is the absolute value of $z^0 \equiv z^D$.

As the above equations indicate, the Gaussian damping factor R(k, z) for *all* large spatial momentum k^j is the same for both the positive and negative frequency parts of the (Two Tier) V Feynman propagator. We are assuming the spatial momentum is real-valued in this discussion. It is also important to note that R(k, z) does not depend on $k^0 = k^D$ (in the V and Y_u Radiation gauges) and thus the integration over k^0 proceeds in the usual way to produce time-ordered positive and negative frequency parts.

The Gaussian exponential factor in *all* spatial coordinates causes the Feynman propagator to be finite and, together with the Gaussian factor in universe particle propagators, causes all perturbation theory calculations when interactions are introduced to be finite as we have seen in Blaha (2017b).

For small momentum much less than M_u then $R(\mathbf{k}, y_1 - y_2) \to 1$ and the Feynman propagator is the "normal" propagator of conventional D-dimensional quantum field theory. For large momentum the corresponding potential approaches r^{D-3} in contrast to the electromagnetic Coulomb potential r^{-1}. The V potential is highly non-singular at large energies.

Thus using Two-Tier Quantum Field Theory we can perform perturbation theory calculations that always yield a finite result.[65] This is not true if conventional Quantum Field is used.[66]

---

[65] In particular, the fermion triangle divergence (anomaly) does not occur in our Two Tier Quantum Field Theory of the fermion sector. Thus there is no requirement for axion-like particles in the Megaverse (or in universes) although the possible existence of this type of particle is not ruled out.

[66] Blaha (2005a) provides a complete discussion of Two-Tier Quantum Field Theory.

41. PseudoQuantum Field Theory

PseudoQuantum Field Theory (and its Quantum Mechanics analogue CQ Mechanics[67]) originates in the need to second quantize in unusual coordinate systems, and in curved space-time coordinate systems. The paper in Appendix 1-A provide a detailed introduction to PseudoQuantum Field Theory.

In this subsection we point out its advantages in a variety of field theory contexts that are relevant for the Unified SuperStandard Theory. The advantages of PseudoQuantum Field Theory are:

1. Quantization in any coordinate system in flat or curved space-times with an invariant definition of asymptotic particle states. An n particle asymptotic state in one coordinate system is a unitarily equivalent n particle asymptotic state in any other coordinate system. Therefore particle number is invariant under change of coordinate system. This is important for the Unified SuperStandard Theory in curved space-times. It is also important for quantization in higher dimensional Euclidean spaces such as the Megaverse. The method was developed in the late 1970's by the author to provide a quantization procedure which supports a unique particle interpretation of states in arbitrary non-static space-times where no global timelike coordinate (Killing vector) exists. PseudoQuantum Field Theory which we developed in a series of books[68] also can be formulated in the Megaverse. Thus we can use it in the Megaverse to implement the Higgs Mechanism to generate particle masses and symmetry breaking.

2. PseudoQuantum Field Theory enables one to define Higgs particle dynamics in such a way that a non-zero vacuum expectation value cleanly separates from the quantum field part of the Higgs fields. This technique can be used in symmetry breaking mechanisms, mass generation, and possible generation of coupling constants as vacuum expectation values.

3. It supports the canonical definition of higher derivative field theories through the use of the Ostrogradski bootstrap. See Appendix B where a fourth order theory of the Strong interaction is defined that has color confinement and a

[67] See Blaha (2016f) for CQ Mechanics, which encompasses both classical mechanics and quantum mechanics, and provides a method of rotating between them. It has applications to transitions between Quantum/Semi-Classical Entanglement, and Quantum/Classical Path Integrals, and Quantum/Classical Chaos.

[68] See Blaha (2017b) for the discussion of the PseudoQuantum field theory formalism for Higgs particles in our Extended Standard Model. See chapter 20 of Blaha (2017b), and earlier books, for a more detailed view than that presented here.

linear r potential. The potential part of this theory was used by the Cornell group to calculate the Charmonium spectrum. (See Blaha (2017b) for details.)

An associated advantage of using PseudoQuantum Field Theory is that it provides for retarded propagators and an Arrow of Time.

41.1 General Case of PseudoQuantization in Differing Coordinate Systems

Papers in Appendix A describe the PseudoQuantization procedure that relates second quantizations in differing coordinate systems. We can epitomize the general concept in the following short example.

Consider the case of a scalar particle in D space-time dimensions that we second quantize in coordinate system denoted 1 with coordinates x based on a timelike Killing vector

$$\varphi(x) = \sum_{\alpha} [\chi_\alpha(x)A_\alpha + \chi_\alpha{}^*(x)A_\alpha{}^\dagger] \qquad (41.1)$$

where the $\chi_\alpha(x)$ are positive frequency with respect to a definition of positive frequency within a universe – following the notation of Appendix A.

Consider now the second quantization of the particle field in a second coordinate system denoted 2 with coordinates y based on a different timelike Megaverse Killing vector

$$\varphi(y) = \sum_{\beta} [\psi_\beta(y)b_\beta + \psi_\beta{}^*(y)b_\beta{}^\dagger] \qquad (41.2)$$

where the $\psi_\beta(y)$ are positive frequency with respect to 2's definition of positive frequency.

Comparing above definitions we see the difference in the definition of the coordinates used in the field expansions as well as the implicit difference in the definitions of positive frequency. To relate the quantizations to each other, we must use the relation between the x and y coordinates:

$$y_i = f_i(x)$$

or, in vector form,

$$y = f(x)$$

for i = 1, 2, … , D. Thus

$$\varphi(f(x)) = \sum_{\beta} [\psi_\beta(f(x))b_\beta + \psi_\beta{}^*(f(x))b_\beta{}^\dagger] \qquad (41.3)$$

Inverting the above equations to obtain the relation of the fourier coefficient operators we see:

$$A_\alpha = \sum_{\beta} [C_{\alpha\beta}\, b_\beta + C'_{\alpha\beta}\, b_\beta{}^\dagger]$$

where $C_{\alpha\beta}$ and $C'_{\alpha\beta}$ are c-number functions of α and β:

$$C_{\alpha\beta} = (\chi_\alpha(x), \varphi(f(x)) \qquad (41.4)$$
$$C'_{\alpha\beta} = (\chi_\alpha{}^*(x), \varphi(f(x))$$

The above equations imply an N particle state in one coordinate system will appear as a superposition of states of various numbers of particles in the other coordinate system IF the standard quantum field theory formulation is used.

TO REMEDY this situation – which we take to be unphysical – we must reformulate quantum field theory using the PseudoQuantum formulation presented Appendix 1-A. The scalar particle case is discussed in Appendix 1-A.

The conclusions of that section, and the sections following it in Appendix 1-A, are:

1. One can define corresponding unitarily equivalent particle states in two quantizations with invariant particle numbers.

2. The fourier coefficient operators of the two quantizations are related by Bogoliubov transformations and are unitarily equivalent.

3. The group of the local Bogoliubov transformations is an infinite tensor product of $SU_{1,1}$ groups.

4. The vacuums of the particles are invariant under Bogoliubov transformations that relate the Megaverse and the universe quantizations.

5. Unitarily equivalent perturbation theories of both quantizations can be defined.

We now consider the case of Two-Tier PseudoQuantization, and then turn to various applications of PseudoQuantization.

41.2 Two-Tier PseudoQuantum Field Theory

The combination of the Two-Tier procedure with the PseudoQuantization procedure leads to a somewhat more complicated situation. In principle, both are required for a Unified SuperStandard Theory in any coordinate system in flat or curved space-times in any number of dimensions. However their direct combination is both complicated and unphysical.

The main purpose of PseudoQuantization is to have particle number invariance under a change of coordinate system. Two-Tier Field Theory ‘cloaks’ each particle in infinite ‘clouds’ of Y^μ quanta as Fig. 40.1 illustrates. We define PseudoQuantization as implementing particle number invariance for ‘bare’ particles without their clouds of Y^μ quanta. Thus an asymptotic particle state of n particles (neglecting its Y^μ quanta cloud) remains a unitarily equivalent n particle state (neglecting its Y^μ quanta cloud) under a change of coordinate system.

To implement this concept we first define quantizations of a particle in coordinate systems without Two-tier quanta. We then 'dress' the quantizations by replacing the coordinates y^μ in each coordinate system with the corresponding Two-Tier coordinates:

$$y^\mu \rightarrow X^\mu(y) = y^\mu + i\, Y^\mu(y)/M_c^2 \qquad (41.5)$$

It appears the most convenient gauge in each coordinate system is the Lorentz gauge:

$$\partial Y^\mu/\partial y^\mu = 0 \qquad (41.6)$$

We now briefly consider the case of a scalar particle PseudoQuantization. We must introduce two fields $\varphi_1(y)$ and $\varphi_2(y)$ with the free fields' lagrangian

$$\mathscr{L}(y) = \partial^\mu\varphi_1\partial_\mu\varphi_2 - \tfrac{1}{2}\,\partial^\mu\varphi_1\partial_\mu\varphi_1 - m^2\varphi_1\varphi_2 + \tfrac{1}{2}\, m^2\, \varphi_1^2 \qquad (41.7)$$

in a coordinate system with coordinates y. Then we arrive at a PseudoQuantum formulation in the coordinate system with coordinates y that is unitarily equivalent to that of a different coordinate system defined a similar manner.

We can replace the c-number coordinates x and y with Two-Tier coordinates of the form

$$X^\mu(y) = y^\mu + i\, Y^\mu(y)/M_c^2$$

and proceed to calculate propagators and perturbation theory diagrams.

Thus we have a straight-forward procedure to unite the PseudoQuantum formalism with Two-Tier coordinates to obtain finite perturbation theory results with unitary equivalence to quantization in other coordinate systems in both flat and curved space-times.

The use of two fields per particle of PseudoQuantum field theory will be seen to part of the applications consider in the remainder of this subsection. We will put aside the consideration of quantizations in other coordinate systems in what follows to keep the presentation as simple as possible.

41.3 PseudoQuantum Higgs Scalar Particle Field Theory in D-dimensional Space-Time

41.3.1 The Enigma of Higgs Particles and the Higgs Mechanism

In our previous work on the Standard Model, and its generalization to The Unified SuperStandard Theory described in a series of books entitled *Physics is Logic* ..., we showed that the fermion spectrum results from Complex Special Relativity, the gauge interactions result from the Reality group, the fermion generations result from the Generation group, and the Theory of Everything results from a combination with Complex General Relativity. The Higgs particles and the Higgs Mechanism were inserted to generate particle masses and symmetry breaking effects.

Whence come Higgs particles? A more fundamental cause has not been suggested until our analysis, which is presented here. So the Higgs sector appeared to be an expedient mechanism to insert much needed symmetry breaking and masses into the theory.

There are a number of peculiarities in the implementation of the Higgs Mechanism:

1. First, it is selective in the sense that some gauge fields have associated Higgs particles and utilize the Higgs Mechanism, and some gauge fields do not have associated Higgs particles. In particular, the ElectroWeak gauge fields, the Generation group gauge fields, the Layer group fields, and the complex gravity Species gauge fields have associated Higgs particles. The strong interaction (gluon) gauge fields do not.

2. The Higgs potentials have a quadratic mass term of the "wrong" sign plus a quartic interaction term, which together, generate non-zero vacuum expectation values. They obviously accomplish their goal. But the source of these potentials, and why they have the same form, is unknown. One expects a fundamental principle should be operative here.

3. One can imagine creating a Higgs microscope at some super-accelerator. Using this microscope in the presence of a (classical) condensate could enable the Uncertainty Principle to be violated. This possibility, in the case of a microscope using electromagnetic fields, was the source of a heuristic argument for the need to quantize the electromagnetic field.[69]

4. The formulation of the Higgs Mechanism uses classical fields under the assumption that a path integral formulation justifies their use. While this may be true, the path integral formulation relies on implicit, unstated boundary conditions that obscure the physics of the quantum field theoretic nature of the mechanism. A direct quantum field theoretic study of the Higgs Mechanism is needed and would further elucidate its character.

5. Scalar fields have a cloud hanging over them that spin ½ fields do not. A spin ½ particle cannot transition to negative energy because there is a filled sea of negative energy particles. No additional particles can fall into the sea due to the Pauli Exclusion Principle that forbids two fermions with the same 4-momentum and quantum numbers. In the case of scalar particles the Pauli Exclusion Principle does not apply and so a *filled* negative energy sea of scalar particles is not possible and positive energy scalar particles can transition to negative energy without hindrance. This problem has been

[69] Heitler (1954) p. 86 provides a good discussion of the need to quantize the electromagnetic field.

"resolved" by an appropriate definition of the scalar particle vacuum to exclude transitions to negative energy. But the rationale for the definition is lacking. Dirac was asked about this issue many years ago. He said he had a solution to the problem. However he did not present it – in keeping with his well-known taciturn nature. So the issue remains an open question.

For the above reasons we will show that a more satisfactory method of achieving the goals of mass generation and symmetry breaking exists.[70] This method relies on a larger Fock space that enables the appearance of a vacuum expectation value for Higgs particles to be understood within a truly quantum framework. More importantly, this method is a consequence the PseudoQuantization procedure described above that enables unitarily equivalent quantizations in different coordinate systems. So a profound fundamental justification for our Higgs boson formulation exists. One major consequence of this approach is the appearance of a local Arrow of Time – a concept that has been a subject of interest for over one hundred years. Another consequence is a rationale for ElectroWeak Higgs bosons and for their absence for the strong (gluon) interaction.

41.3.2 PseudoQuantization of Scalar Particles

We now consider the PseudoQuantization[71] of a scalar particle field that will become a Higgs particle with a non-zero vacuum expectation value.[72] We begin by defining two fields that correspond to the scalar particle: $\varphi_1(x)$ and $\varphi_2(x)$.[73] These fields will be assumed to have the equal time commutators

$$\begin{aligned}
&[\varphi_i(x), \pi_j(y)] = i(1 - \delta_{ij})\delta^3(\mathbf{x} - \mathbf{y}) \qquad (41.8)\\
&[\varphi_i(x), \varphi_j(y)] = 0\\
&[\pi_i(x), \pi_j(y)] = 0
\end{aligned}$$

where δ_{ij} is the Kronecker δ and where $\pi_i(x)$ is the canonically conjugate momentum to $\varphi_i(x)$. The fields $\varphi_1(x)$ and $\pi_1(y)$ will be observable classical fields. The fields $\varphi_2(x)$ and $\pi_2(y)$ will not be observables so that $\varphi_1(x)$ and $\pi_1(y)$ can both be sharp on the set of physical states.

We now specify the lagrangian density for a scalar Klein-Gordon particle:

---

[70] In the Extended Standard Model of Blaha (2015a) we have shown that the basic particles have a mass, the Landauer mass, so that the theory is symmetry violating from the very start. We have also shown that our Two-Tier formalism for quantum field theories always yields finite results in perturbation theory calculations – making the renormalization approach of t'Hooft and others, which relied on initially massless gauge fields, unnecessary.

[71] PseudoQuantization in a D-dimensional space-time is described in Blaha (2017c). This discussion is relevant to PseudoQuantization in the Megaverse, or in other universes.

[72] Much of this section appears in Blaha (2016c), and earlier books, as well as in S. Blaha, Phys. Rev. **D17**, 994 (1978). The case of fermion PseudoQuantization is also discussed in Appendix A – S. Blaha, Il Nuovo Cimento **49A**, 35 (1979).

[73] The subscripts on the fields are not gauge symmetry indices but simply identifiers distinguishing the fields from each other.

$$\mathcal{L} = \partial\varphi_1/\partial x_\mu \partial\varphi_2/\partial x^\mu \tag{41.9}$$

with hamiltonian density

$$\mathcal{H} = \pi_1\, \pi_2 + \partial\varphi_1/\partial x_i \partial\varphi_2/\partial x^i$$

where i labels spatial coordinates, and $\pi_1 = \partial\varphi_2/\partial t$ and $\pi_2 = \partial\varphi_1/\partial t$. The lagrangian $\mathcal{L}$ is without a potential or mass term.

The lagrangian and hamiltonian for a massive scalar particle in this formalism are

$$\mathcal{L} = \partial\varphi_1/\partial x_\mu \partial\varphi_2/\partial x^\mu - m^2\, \varphi_1\varphi_2 \tag{41.10}$$

with hamiltonian density

$$\mathcal{H} = \pi_1\, \pi_2 + \partial\varphi_1/\partial x_i \partial\varphi_2/\partial x^i + m^2\, \varphi_1\varphi_2$$

The fields can be fourier expanded in terms of creation and annihilation operators:

$$\varphi_i(\mathbf{x}, t) = \int d^3k\, [a_i(k)f_k(x) + a_i^\dagger(k)f_k^*(x)] \tag{41.11}$$

for i = 1, 2 where

$$f_k(x) = e^{-ik\bullet x}/(2\omega_k(2\pi)^3)^{1/2}$$

with $\omega_k = |\mathbf{k}|$.

The creation and annihilation operators satisfy the commutation relations:

$$[a_i(k), a_j^\dagger(k')] = (1 - \delta_{ij})\delta^3(\mathbf{k} - \mathbf{k'})$$
$$[a_i(k), a_j(k')] = 0$$
$$[a_i^\dagger(k), a_j^\dagger(k')] = 0$$

for i, j = 1, 2.

In this formulation the defining properties of a physical state are:

$$\varphi_1(x)|\Phi, \Pi> = \Phi(x)|\Phi, \Pi> \tag{41.12}$$
$$\pi_1(x)|\Phi, \Pi> = \Pi(x)|\Phi, \Pi>$$

where $\Phi(x)$ and $\Pi(x)$ are sharp on the states and thus classical fields with

$$\Phi(\mathbf{x}, t) = \int d^3k\, [\alpha(k)f_k(x) + \alpha^*(k)f_k^*(x)] \tag{41.13}$$

and correspondingly for $\Pi(x)$.

41.3.3 Vacuum States for Scalar (Higgs) Particles with Non-Zero Vacuum Expectation Values

When we implement the mass mechanism, Φ is constant. We can define a set of states

$$a_1(k)|\alpha> = \alpha(k)|\alpha>$$
$$a_1^\dagger(k)|\alpha> = \alpha^*(k)|\alpha>$$

and correspondingly a set of coherent states

$$|\alpha\rangle = C\exp\{\int d^3k\ [\alpha(k)a_2^\dagger(k) + \alpha^*(k)a_2(k)]\}|0\rangle \quad (41.14)$$

where C is a normalization constant and where the vacuum state $|0\rangle$ satisfies

$$a_1(k)|0\rangle = a_1^\dagger(k)|0\rangle = 0 \quad (41.15)$$

$$a_2(k)|0\rangle \neq 0 \qquad\qquad a_2^\dagger(k)|0\rangle \neq 0$$

The dual vacuum state satisfies

$$\langle 0|a_2(k) = \langle 0|a_2^\dagger(k) = 0$$

$$\langle 0|a_1(k) \neq 0 \qquad\qquad \langle 0|a_1^\dagger(k) \neq 0$$

With this coherent state formalism, which gives purely classical fields and yet also has quantum fields through the use of φ_2 and its creation and annihilation operators, we now have the machinery to define a mass mechanism without the introduction of a potential whose origin can only be described as dubious.

For we can define a coherent state for some k as

$$|\Phi, \Pi\rangle = C\exp\{[(2\pi)^3\omega_k/2]^{1/2}\Phi[a_2^\dagger(k) + a_2(k)]\}|0\rangle \quad (41.16)$$

where C is a normalization constant, which yields a non-zero vacuum expectation value:

$$\varphi_1(x)|\Phi, \Pi\rangle = \Phi|\ \Phi, \Pi\rangle \quad (41.17)$$

where Φ is a constant. Evaluating a fermion interaction term we find a mass term emerges[74]

$$\psi\ (\varphi_1 + \varphi_2)\psi\ \rightarrow \bar{\psi}(\Phi + \varphi_2)\psi \quad (41.18)$$

It generates a mass for an interaction with a gauge field of the form

$$A^\mu(\varphi_1 + \varphi_2)^2A_\mu \rightarrow A^\mu(\Phi + \varphi_2)^2A_\mu \quad (41.19)$$

It also yields a quantum field theoretic interaction that would result in the production of ElectroWeak particles from these scalar fields. The production of Higgs particles that decay into ElectroWeak gauge particles has recently been found at CERN.

The present formalism provides a clean way to separate the vacuum expectation value of a scalar particle from its quantum field part in contrast to the Higgs Mechanism where one has to separate a Higgs field into parts manually.

---

[74] When matrix elements with a "vacuum state" are taken.

41.3.4 Interpretation of Negative Energy Scalar Particle States

As we noted earlier, scalar particle physics has the problem of no barrier to the decay of positive energy states to negative energy states due to the absence of a Pauli Exclusion Principle for bosons. The PseudoQuantization procedure that we developed in 1978 and describe here allows negative energy states as one would physically expect and raises the possibility of disastrous particle decays to negative energy. The above equations show that negative energy states are possible in this theory.

However they also show that combined positive and negative energy boson states can be interpreted as classical field states. In addition, the ability of any number of boson particles to have the same 4-momentum and quantum numbers shows that a *macroscopic* classical scalar field state can be constructed.

Thus we can view states containing negative energy particles as classical field states and thus solve[75] *the issue of interpreting negative energy particle states – a more satisfactory approach than the standard quantization procedure does – with due respect to Professor Dirac.*

We note that macroscopic many particle fermion states can only have one particle in any mode unlike bosons. Therefore we cannot use this formalism to create macroscopic classical fermion field states.[76] And the filled Dirac sea of negative energy fermions precludes the transition of a positive energy Dirac fermion to a negative energy state. *Thus there is a certain complementarity between fermions that cannot become classical fields but have a filled sea precluding decays to negative energy states, and bosons that can become classical fields but support decays to negative energy states.*

41.3.5 Contrast with Conventional Second Quantization of Scalar Particles

The PseudoQuantization procedure followed here uses different boundary conditions than the usual scalar particle quantization procedure. The essence of the difference is embodied in a comparison of the definition of the vacuum above and the definition of the conventional second quantized field vacuum:

$$a|0\rangle = 0 \qquad\qquad \text{Conventional Approach}$$
$$a^\dagger|0\rangle \neq 0$$

In the conventional approach the creation of negative energy boson states is eliminated *ab initio* whereas in our approach it is allowed in order to support classical field states with non-zero vacuum expectation values that are a form of classical field. While one cannot discredit the conventional choice for conventional scalar fields, one can see that our approach yields a physically more important result – particularly for Higgs fields – because it leads to an Arrow of Time *locally* – an important feature of physical

[75] Also a boson that has no interactions cannot transition from to a positive energy state to a negative energy state due to conservation of energy.

[76] However we can create PseudoQuantum fermion states. See S. Blaha, Phys. Rev. **D17**, 994 (1978) (reproduced in Appendix I) and references therein to earlier papers by the author.

phenomena that has been a subject of much discussion and dispute. One can say that the conventional approach sweeps the issue "under the rug" rather than seeking a deeper justification – differing from Dirac's implied notion that the issue merited attention. We will discuss the "Arrow of Time" within the framework of our PseudoQuantization approach later.

41.3.6 Why Inertial Reference Frames are Special

The great physicists of the early 20th century raised numerous questions about Special Relativity after Einstein and Poincarè's discovery. Prominent among them was the question of why inertial reference frames are of especial importance in Special Relativity, and afterwards in General Relativity.

It appears that our formulation of the mass generation mechanism sheds significant light on the reason for the special prominence of inertial frames. Earlier we considered the case of a massless PseudoQuantized scalar. We now consider massive scalars since experiments at CERN have apparently discovered a Higgs particle with a 125 GeV/c mass. The above equations describe a massive scalar particle. If the scalar is massive, then the "vacuum" state that yields a non-zero expectation value must change to

$$|\Phi, \Pi\rangle = C\exp\{(2\pi)^3 m/2]^{1/2}[a_2^\dagger(\mathbf{0},m) + a_2(\mathbf{0},m)]\}|0\rangle \qquad (41.20)$$

to have operators for a particle of mass m in its rest frame. Then, having established this preferred frame for a Higgs particle, in The Unified SuperStandard Theory, and requiring that invariant intervals

$$ds^2 = dt^2 - d\mathbf{x}^2 \quad \text{(in rectangular coordinates)}$$

are unchanged by a (complex or real) Lorentz transformation, we find that inertial reference frames are singled out as "special" in the sense that they are the only accessible reference frames that can be generated by a Lorentz boost/transformation from the Higgs particle rest frame. *The Higgs particle vacuum state singles out the class of inertial reference frames.*

Thus Higgs particles play a central role in establishing the basis of physical reality.

41.3.7 PseudoQuantization Reveals More Physical Consequences than the Higgs Mechanism of Scalar Particles

Earlier we pointed out that our PseudoQuantization theory of Higgs particles reveals more physical consequences than the conventional approach, which implements the Higgs Mechanism by simply using a potential term that has a minimum at a non-zero vacuum expectation value. This section shows the major results of a properly implemented mechanism. We find a better explanation of the negative energy state problem of boson field theories. We find a local arrow of time that explains the direction of time that we, and all of nature, experiences. We find the reason why inertial reference frames have a special physical significance – a result long sought by physicists.

In addition we will see in chapter 11 that real gauge fields should have an associated Higgs particle, while necessarily complex gauge fields (the Strong interaction gauge field in The Unified SuperStandard Theory) do not have an associated gauge field. These results correspond to experimental reality.

41.3.8 The T Invariance Issues of Our PseudoQuantized Scalar Particle Theory

The PseudoQuantized scalar particle hamiltonian equations are invariant under time reversal $t \rightarrow t' = -t$. The 'new' vacuum states defined above break the time reversal invariance of the theory resulting in retarded particle propagators.

The hamiltonian equations

$$[H, \varphi_1(\mathbf{x}, t)] = -i\partial\varphi_1/\partial t \qquad (41.21)$$
$$[H, \varphi_2(\mathbf{x}, t)] = -i\partial\varphi_2/\partial t$$

are invariant under time reversal. If we define a time reversal operator transformation U then the time reversed equations are

$$[UHU^{-1}, \varphi_1(\mathbf{x}, -t)] = +i\partial\varphi_1(\mathbf{x}, -t)/\partial(-t)$$
$$[UHU^{-1}, \varphi_2(\mathbf{x}, -t)] = +i\partial\varphi_2(\mathbf{x}, -t)/\partial (-t)$$

The operator U, which is unitary, transforms H into **–H**. This operation is legal because the hamiltonian – in this case – is not positive definite and admits negative energy states.[77] Thus

$$[H, \varphi_1(\mathbf{x}, -t)] = -i\partial\varphi_1(\mathbf{x}, -t)/\partial (-t)$$
$$[H, \varphi_2(\mathbf{x}, -t)] = -i\partial\varphi_2(\mathbf{x}, -t)/\partial (-t)$$

and the time reversal invariance of the equations of motion is established for this case.

Time reversal invariance is broken by our choice of vacuum states. This choice is necessary to obtain classical field states as we showed earlier. A demonstration of the time reversal symmetry breaking is presented later where we show theory has retarded propagators for particle propagation to and from asymptotic states.

Within the interaction region the particle propagators are the sum of retarded and advanced parts that combine to yield principle value propagators – not Feynman propagators. Many years ago Feynman and Wheeler championed principle value propagators for electrodynamics to obtain an action-at-a distance theory of Quantum Electrodynamics. While their theory, and ours, differ from the standard quantum field theory approach there is no reason to view them as faulty, or having serious physical defects. The only question is whether nature chooses conventional quantum field theory

[77] Unlike the usual case of second quantized Klein-Gordon quantum field theory.

or PseudoQuantized quantum field theory. In our case the need for a classical scalar particle non-zero vacuum expectation value strongly motivates our choice of pseudoquantized Higgs particles.

41.3.9 Retarded Propagators for Our Quantized Higgs Particles

In the previous section we pointed out that our PseudoQuantization Higgs theory has an arrow of time due to its boundary conditions as expressed by its definition of the vacuum state and its dual. In this section we will show that the theory uses retarded propagators for propagation to and from the interaction region to asymptotic in-states and out-states. Within an interaction region the theory uses half-retarded – half-advanced propagators. We discuss aspects of the perturbation theory and propagators of our scalar particles in this chapter.

First we note that in-states at $t = -\infty$ are composed of superpositions of $a_2(k)$ and $a_2^\dagger(k)$ creation and annihilation operators:

$$a_2(k)|0\rangle \neq 0 \qquad\qquad a_2^\dagger(k)|0\rangle \neq 0$$

while the out-states composed of superpositions of $a_1(k)$ and $a_1^\dagger(k)$ creation and annihilation operators:

$$\langle 0|a_1(k) \neq 0 \qquad\qquad \langle 0|a_1^\dagger(k) \neq 0$$

Consequently when in-state particles (x_1) propagate into the interaction region (x_2) the relevant propagators are retarded propagators with the form

$$\begin{aligned} G_{in}(x_2, x_1) &= \langle 0|T(\varphi_{1\,in}(x_2), \varphi_{2\,in}(x_1))|0\rangle \\ &= \theta(x_{20} - x_{10})\langle 0|[\varphi_{1\,in}(x_2), \varphi_{2\,in}(x_1)]\,|0\rangle \end{aligned} \tag{41.22}$$

This is a manifestly retarded propagator. The choice of vacuums clearly results in a time asymmetry giving a retarded propagation reflecting the familiar Arrow of Time.

A similar situation prevails for propagation to out-states (x_3) from the interaction (x_2) region:

$$\begin{aligned} G_{out}(x_3, x_2) &= \langle 0|T(\varphi_{1\,out}(x_3), \varphi_{2\,out}(x_2))|0\rangle \\ &= \theta(x_{30} - x_{20})\langle 0|[\varphi_{1\,out}(x_3), \varphi_{2\,out}(x_2)]\,|0\rangle \end{aligned} \tag{41.23}$$

Within the interaction region the Higgs particles have principle value propagators.

Thus we find PseudoQuantized Higgs particles embody a local Arrow of Time. The locality of the Arrow of Time is embodied in all the particles that interact with the Higgs particle. Since the mass of *every* particle – bosons and fermions – has a Higgs contribution, and thus *every* particle interacts with the Higgs particles, the Arrow of Time permeates The Unified SuperStandard Theory as well as the more familiar Standard Model known from experiment.

41.3.10 The Local Arrow of Time

In the *Physics is Logic* series of monographs we saw that complex coordinates led to the form of the fermion spectrum, that the mapping of complex coordinates to real-valued coordinates yielded the Reality group and The Unified SuperStandard Theory gauge interactions, that Complex General Relativity led to Higgs particles that were directly united with elementary particle masses and gave us the equality of inertial mass and gravitational mass. Later we will see the reduction of complex gauge fields to real gauge fields explains the appearance of Higgs fields in The Unified SuperStandard Theory.

The PseudoQuantization procedure leads to retarded Higgs field propagators and thence to a *local* arrow of time. Many arguments have been put forward over the past hundred plus years for the Arrow of Time. Many arguments based on Statistical Mechanics, Entropy, and Boltzmann's statistical atomic theory have suggested the Arrow of Time is a global statistical consequence. This view seems to contradict the results of elementary particle experiments where a *local* Arrow of Time is evident.

Our rationale for the Arrow of Time begins with retarded Higgs fields. Then we note that Higgs field quantum interactions appear for all fermions and gauge particles. Thus all particle interactions are imbued with an Arrow of Time. Particles united to form macroscopic matter inherit their combined Arrows of Time producing the global Arrow of Time we experience.

Thus our PseudoQuantization approach offers a more satisfactory solution of the origin of the Arrow of Time.

It is remarkable that complex quantities – coordinates and fields – through the Higgs phenomena that we have considered, lead to the equality of inertial mass and gravitational mass, and an Arrow of Time. This unity of mass and time phenomena may reflect the deeper fact that we can have no practical Arrow of Time if all particles were massless, for particle dynamics at light speed would then be pointless. This view has been expressed by DeWitt, Unruh, and others who have pointed out that, physically, time is meaningful and measurable only if masses exist; the larger the mass, the more accurate the time measurement in principle.[78]

41.3.11 Space-Time Dependent Particle Masses

It is possible that the ultimate Unified SuperStandard Theory has masses that evolve with time and may also be spatially varying – different values in different parts of the universe. Presently there is no decisive evidence for this possibility although astrophysical studies continue. In this section we will describe the mechanism for space-time dependent masses.

Consider a classical field (time and spatially varying):

$$\Phi(\mathbf{x}, t) = \int d^3k\, [\alpha(k)f_k(x) + \alpha^*(k)f_k{}^*(x)] \qquad (41.24)$$

[78] No mass, no clock; no clock, no physical time. See Blaha (2015a) pp. 368-371 for a discussion including comments by DeWitt and Unruh.

If we define the coherent vacuum state

$$|\alpha> = C \exp\{\int d^3k\, [\alpha(k)a_2^{\dagger}(k) + \alpha^*(k)a_2(k)]\}|0> \qquad (41.25)$$

then

$$\varphi_1(x)|\Phi, \Pi> = \Phi(x)|\Phi, \Pi>$$
$$\pi_1(x)|\Phi, \Pi> = \Pi(x)|\Phi, \Pi>$$

where

$$\varphi_i(\mathbf{x}, t) = \int d^3k\, [a_i(k)f_k(x) + a_i^{\dagger}(k)f_k{}^*(x)] \qquad (41.26)$$

for i = 1, 2 and where

$$f_k(x) = e^{-ik\cdot x}/(2\omega_k(2\pi)^3)^{1/2}$$

with ω_k equal to the energy.

41.3.12 Inertial Mass Equals Gravitational Mass

From the days of Newton through Einstein[79] to the present the equality of gravitational mass and inertial mass has been a topic of interest. Mach, who played an important role, in this ongoing discussion, thought distant masses in the universe were the source of the equality. However the origin of the equality, which has been shown experimentally to very high accuracy, remained uncertain until the *Physics is Logic* series of books, in which we showed the interconnection of the Unified SuperStandard Theory and Complex Gravitation via Higgs generated masses that united gravitational and inertial mass.

In Blaha (2016h) we showed that a Complex General Relativity transformation can be factored into the product of a complex-valued transformation and a real-valued General Coordinate transformation. The set of complex valued transformations form a U(4) group that we called the General Coordinate Reality group. Later we will define the Internal Symmetry Species Group as the corresponding analogue. The Species Group has gauge fields that undergo spontaneous symmetry breaking and generate contributions to all fermion masses.

Since fermion field masses are now sums of ElectroWeak Higgs contributions, Generation group Higgs contributions, Layer group Higgs contributions, and Species group contributions, and since the gravitational Higgs fields appear in all fermion masses, the equality of inertial and gravitational mass is proven. The gravitational Higgs particles' equations depend, in part, on the gravitational field by Blaha (2016h) and so set the mass scale of gravitational mass, and thereby of all Higgs mass contributions. They set the scale of inertial masses equal to the scale of gravitational masses. ***Since an expression cannot mix mass scales, the gravitational mass scale must be the same as the inertial mass scale. Inertial Mass equals gravitational mass.***

---

[79] For example, Einstein and Grossman in 1913 stated, "The theory herein described originates in the conviction that the proportionality between the inertial and gravitational mass of a body is an exact law of nature that must be expressed as a foundation principle of theoretical physics."

We have established the equality of inertial and gravitational mass at the short distance quantum level. In our view, this explanation is far more satisfying than basing the equality on a combination of large distance phenomena and quantum phenomena. As Einstein and Weyl have pointed out, all fundamental physics phenomena should be based on a local theory. Complex Gravity as we have constructed it, combined with the Unified SuperStandard Theory, furnishes a completely local basic Theory of Everything.

The equation above contains a coherent state $|\alpha>$ for a time and spatially varying mass. The above equations can be generalized to the case of multiple space-time varying masses.[80]

$$|\Phi_1,\Phi_2, \ldots ,\Phi_n;\Pi_1,\Pi_2, \ldots ,\Pi_n> = C \prod_{i=1}^{n} \exp\left\{\int d^3k \, [\alpha_i(k)a_{2i}^\dagger(k) + \alpha_i^*(k)a_{2i}(k)]\right\}|0> \quad (41.27)$$

Then all n mass vacuum expectation values are space-time dependent:

$$\varphi_{1i}(x) \,|\, \Phi_1, \Phi_2, \ldots , \Phi_n; \Pi_1, \Pi_2, \ldots , \Pi_n> = \Phi_i(x) \,|\, \Phi_1, \Phi_2, \ldots , \Phi_n; \Pi_1, \Pi_2, \ldots , \Pi_n> \quad (41.28)$$

Thus our formalism can accommodate space-time varying masses should they be found in the Cosmos.

41.3.13 Benefits of the PseudoQuantization Method

In this book, and in earlier work, we showed that a more physically satisfactory method for avoiding the negative energy state problem exists. This method relies on the use of a larger Fock space in which negative energy states (or partially negative energy states) are interpreted as states containing classical fields or a mix of classical fields and individual boson particles. This approach resolves the negative energy boson issue and provides a common framework for boson particles and classical boson fields.

One consequence of the PseudoQuantization method is that it enables the appearance of a vacuum expectation value for Higgs particles (a constant classical field) to be understood within a truly quantum framework. Another major consequence of this approach is the appearance of a *local* Arrow of Time due to the Higgs mass generation mechanism – a concept that has been a subject of interest for over one hundred years. A macroscopic arrow of time is often described as a statistical result. But our approach yields an arrow of time at the single particle level.

The conventional approach to boson field quantization sweeps these issues "under the rug" rather than seeking a deeper justification. It differs from Dirac's implied notion that the issue merited attention.

Another important consequence of the PseudoQuantization method is that it singles out inertial reference frames when applied to the case of Higgs particles.

---

[80] The "vacuum" state $|0>$ also implicitly has factors for the vacuum expectation values used for fields that give masses to fermions and vector bosons as described in Blaha (2016h).

Yet another more subtle consequence of boson PseudoQuantization is that it provides a rationale/explanation for the presence of ElectroWeak Higgs bosons, *and for their absence for the strong (gluon) interactions. The question of why there are no strong interaction Higgs bosons has not been previously considered to the best of this author's knowledge.*

41.4 Two-Tier Formulation and PseudoQuantization in the Megaverse

These subjects are considered in Blaha (2020c) in chapter 68 as well as in earlier books by the author.

Appendix B. Complex General Relativity & Dimension – Index Connection From Chapters 50 and 51 of Blaha (2020c)

The chapters in this appendix are from Blaha (2020c). Much of those chapters also appeared in Blaha (2016h) and (2017a). They show the connection between dimensions (coordinates) and indices in *four* dimension Complex General Relativity.

50. Complex General Relativity Reformulated

50.1 Tetrad (Vierbein) Formalism

The *vierbein* formalism begins with the Equivalence Principle that allows us to define an inertial coordinate system in the neighborhood of any point Z in space-time. We will use the notation $\varsigma^\alpha(Z)$ to denote the inertial coordinates at Z. We define a tetrad or vierbein as

$$v^\alpha{}_\mu(x) = (\partial\varsigma^\alpha(x)/\partial x^\mu)_{x=Z} \tag{50.1}$$

and, in a neighborhood of Z, we can invert the relation between ς and x to define an inverse

$$w^\mu{}_\alpha(x) = (\partial x^\mu(\varsigma)/\partial\varsigma^\alpha)_{x=X} \tag{50.2}$$

such that

$$\begin{aligned} w^\mu{}_\alpha(x)v^\alpha{}_\nu(x) &= \delta^\mu{}_\nu \\ w^\mu{}_\beta(x)v^\alpha{}_\mu(x) &= \delta^\alpha{}_\beta \end{aligned} \tag{50.3}$$

In real General Relativity all *tetrads* are real-valued. In Complex General Relativity a *tetrad* $v^\alpha{}_\mu(x)$ is complex-valued.

The metric at a curved space-time point X is defined in terms of *tetrads* as

$$\begin{aligned} g_{\rho\sigma}(x) &= \eta_{\alpha\beta}\, v^\alpha{}_\rho(x)v^\beta{}_\sigma(x) \\ g^{\rho\sigma}(x) &= \eta^{\alpha\beta}\, w^\rho{}_\alpha(x)w^\sigma{}_\beta(x) \end{aligned} \tag{50.4}$$

The inverse of a *tetrad* transformation can also be expressed as

$$w_\beta{}^\nu(x) = v_\beta{}^\nu(x) = \eta_{\beta\alpha}g^{\nu\mu}(x)v^\alpha{}_\mu(x)$$

Then a *tetrad* and its inverse satisfy

$$v^\alpha{}_\mu(x)v_\beta{}^\mu(x) = \delta^\alpha{}_\beta \tag{50.5}$$

and

$$v^\alpha{}_\mu(x)v_\alpha{}^\nu(x) = \delta^\nu{}_\mu$$

There are two general types of space-time transformations that can be performed on a tetrad.

1. A complex-valued (possibly real-valued) General Relativistic coordinate transformation:

$$v'^{\alpha}{}_{\mu}(x) = \partial x^{\nu}/\partial x'^{\mu}\, v^{\alpha}{}_{\nu}(x)$$

2. A complex-valued, local *Lorentzian transformation*

$$v'^{\beta}{}_{\mu}(x) = \Lambda(x)^{\beta}{}_{\alpha}\, v^{\alpha}{}_{\mu}(x)$$

where $\Lambda(x)^{\beta}{}_{\alpha}$ is an element of a subset of the local Complex Lorentz Group.

The local Lorentzian transformations $\Lambda(x)^{\beta}{}_{\alpha}$ consist of local Lorentz transformations that are real-valued, and complex-valued Lorentz transformations. Both types of transformations satisfy the orthogonality condition:

$$\eta_{\alpha\beta}\Lambda^{\alpha}{}_{\rho}(x)\Lambda^{\beta}{}_{\sigma}(x) = \eta_{\rho\sigma} \tag{50.6}$$

Thus the *tetrad* partakes of both local (position dependent) General Relativistic transformations and local Lorentzian transformations.

50.2 Complex General Relativistic Transformations

The General Relativistic Reality group interaction emerges from complex General Relativistic transformations. We can separate elements of the set of all complex General Coordinate transformations into a product of two factors: a real-valued General Coordinate transformation and a complex-valued General Coordinate transformation. The set of complex factors can be further factored into those that satisfy

$$\Lambda(\omega, \mathbf{u})^{T}G\Lambda(\omega, \mathbf{u}) = G \tag{50.7}$$

and those that do not. We then see that the set of those that do not satisfy the above equation form a curved space representation of the U(4) group under 'multiplication' of transformations.

The elements of the set of real and complex General Coordinate transformations whose flat complex space-time limit satisfy the above equation form the elements of the Complex Lorentz group.[81]

We thus find the set of all 4-dimensional complex, curved space General coordinate transformations can be visualized as in Fig. 50.1. The next section describes the interplay of the three parts displayed in Fig. 50.1.

---

[81] It is this part of curved space-time General Relativity that becomes the flat space-time Complex Lorentz group, which leads to the SU(3)⊗SU(2)⊗U(1)⊗SU(2)⊗U(1)⊗SU(3) Standard Model group.

50.3 Structure of Complex General Coordinate Transformations

Complex General Coordinate transformations can be uniquely factored into products of two terms, which will later be further factored into three factors. They have the form

$$\partial x''^{\nu}(x)/\partial x^{\mu} = U(x'')^{\nu}{}_{\beta}\, \partial x'^{\beta}(x)/\partial x^{\mu} \tag{50.8}$$

where

$$x''^{\nu}(x) = U(x'')^{\nu}{}_{\beta} x'^{\beta}$$
$$x'^{\mu}(x) = U^{-1\mu}{}_{b}(x'')\, x''^{b}$$

where $U(x')^{\nu}{}_{\beta}$ is complex and where $\partial x'^{\beta}(x)/\partial x^{\mu}$ is a purely real General Coordinate transformation.

We define

$$U(x'')^{\mu}{}_{\nu} = w^{\mu}{}_{a}(x'')\left[\exp(i \textstyle\sum_k g_k \Phi_k(x'')\tau_k)\right]^{a}{}_{b}\, v^{b}{}_{\nu}(x'') \tag{50.9}$$
$$U^{-1}(x'')^{\mu}{}_{\nu} = w^{\mu}{}_{a}(x'')\left[\exp(-i \textstyle\sum_k g_k \Phi_k(x'')\tau_k)\right]^{a}{}_{b}\, v^{b}{}_{\nu}(x'')$$

where the constants g_k are real, and Φ_k and τ_k are hermitean. The uniqueness of the factorization follows from the Reality group (and U(4)) property that any complex 4-vector can be uniquely mapped to any specified real 4-vector.

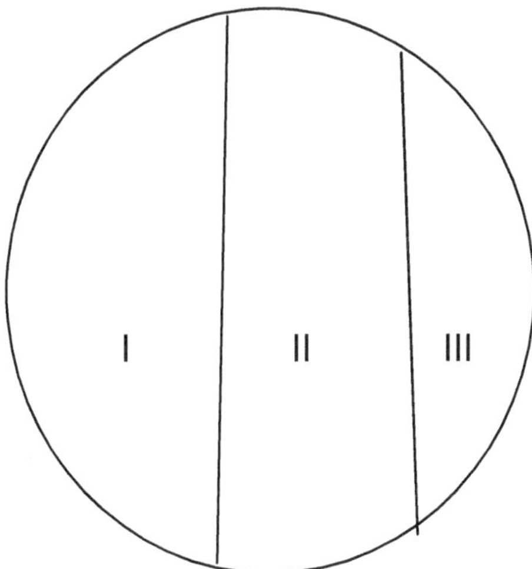

Figure 50.1. A visualization of the set of General Coordinate transformations separated into real-valued General coordinate transformations (part I), complex transformations that satisfy $\Lambda(\omega, u)^{T} G \Lambda(\omega, u) = G$ (part II), and complex transformations that do not satisfy $\Lambda(\omega, u)^{T} G \Lambda(\omega, u) = G$ (part III). Part I and part II combine in the limit of flat space-time to form the Complex Lorentz group. Parts II and III elements form a U(4) group that we call the General Relativistic Reality group.

Given the factorization above it becomes possible to separate the affine connection correspondingly.

50.4 Complex Affine Connection – General Relativistic Reality Group

The structure of a complex general coordinate transformation enables us to calculate its affine connection for later use in determining the covariant derivative, and the dynamic equations. First the transformation to the real-valued x' coordinates from inertial coordinates is

$$\Gamma^{\sigma}{}_{\lambda\mu}(x') = \partial x'^{\sigma}/\partial\varsigma^{\rho}\, \partial^{2}\varsigma^{\rho}/\partial x'^{\lambda}\partial x'^{\mu} \qquad (50.10)$$

Next the Reality group transformation has the affine connection

$$\Gamma^{\sigma}{}_{\lambda\mu}(x'') = \partial x''^{\sigma}/\partial\varsigma^{\rho}\, \partial^{2}\varsigma^{\rho}/\partial x''^{\lambda}\partial x''^{\mu}$$

which can be re-expressed as

$$\begin{aligned}\Gamma^{\sigma}{}_{\lambda\mu}(x'') &= \partial x''^{\sigma}/\partial x'^{\beta}\, \partial x'^{\beta}(\varsigma)/\partial\varsigma^{\rho}\, \partial/\partial x''^{\mu}[\partial\varsigma^{\rho}/\partial x'^{\alpha}\, \partial x'^{\alpha}/\partial x''^{\lambda}] \\ &= \partial x''^{\sigma}/\partial x'^{\beta}\, \partial x'^{\alpha}/\partial x''^{\lambda}\, \partial x'^{\gamma}/\partial x''^{\mu}\, \Gamma^{\beta}{}_{\alpha\gamma}(x') + \partial x''^{\sigma}/\partial x'^{\beta}\, \partial^{2}x'^{\beta}/\partial x''^{\lambda}\partial x''^{\mu}\end{aligned} \qquad (50.11)$$

Next substituting the General Relativistic Reality group transformation

$$\begin{aligned}x''^{\nu}(x) &= U(x'')^{\nu}{}_{\beta}x'^{\beta} \\ x'^{\mu}(x) &= U^{-1}(x'')^{\mu}{}_{\beta}\, x''^{\beta}\end{aligned}$$

together with

$$\partial x''^{\sigma}/\partial x'^{\beta} = \partial[U(x'')^{\sigma}{}_{\alpha}x'^{\alpha}]/\partial x'^{\beta} = U(x'')^{\sigma}{}_{\beta} + x'^{\alpha}\, \partial U(x'')^{\sigma}{}_{\alpha}/\partial x'^{\beta}$$

$$\partial x'^{\sigma}/\partial x''^{\beta} = \partial[U^{-1}(x'')^{\sigma}{}_{\alpha}x''^{\alpha}]/\partial x''^{\beta} = U^{-1}(x'')^{\sigma}{}_{\beta} + x''^{\alpha}\, \partial U^{-1}(x'')^{\sigma}{}_{\alpha}/\partial x''^{\beta}$$

we find the second term above is the Reality fields affine connection

$$\Gamma_{R}{}^{\sigma}{}_{\lambda\mu}(x'') = \partial[U(x'')^{\sigma}{}_{\alpha}x'^{\alpha}]/\partial x'^{\beta}\, \partial\{\partial[U^{-1}(x'')^{\beta}{}_{\alpha}x''^{\alpha}]/\partial x''^{\lambda}\}/\partial x''^{\mu}$$

and so we find the affine connections are approximately additive. Thus approximately

$$\Gamma^{\sigma}{}_{\lambda\mu}(x'') = \Gamma_{GR}{}^{\sigma}{}_{\lambda\mu}(x') + \Gamma_{R}{}^{\sigma}{}_{\lambda\mu}(x'')$$

if $x''^{\sigma} \simeq x'^{\sigma}$.

A complex transformation of types II and III in Fig. 50.1 has the form:

$$\begin{aligned}U(x'')^{\mu}{}_{\nu} &= w^{\mu}{}_{a}(x'')[\exp(i\textstyle\sum_{k}\Phi_{k}(x'')\tau_{k})]^{a}{}_{b}\, 1^{b}{}_{\nu}(x'') \\ U^{-1}(x'')^{\mu}{}_{\nu} &= w^{\mu}{}_{a}(x'')[\exp(-i\textstyle\sum_{k}\Phi_{k}(x'')\tau_{k})]^{a}{}_{b}\, 1^{b}{}_{\nu}(x'')\end{aligned}$$

where τ_k is a U(4) generator matrix. Its infinitesimal transformation is approximately

$$U(x'')^{\nu}{}_{\beta} \approx \delta^{\nu}{}_{\beta} + i \sum_k \Phi_k(x'')[\tau_k]^{\nu}{}_{\beta} \quad (50.12)$$
$$U^{-1}(x'')^{\nu}{}_{\beta} \approx \delta^{\nu}{}_{\beta} - i \sum_k \Phi_k(x'')[\tau_k]^{\nu}{}_{\beta}$$

using the *vierbein* flat space-time limits

$$w^{\mu}{}_{a}(x'') \approx \delta^{\mu}{}_{a}$$
$$1^{b}{}_{\nu}(x'') \approx \delta^{b}{}_{\nu}$$

where

$$\Phi_k(x) = \int^{x} dy_{\lambda}\, A_{Rk}{}^{\lambda}(y) \quad (50.13)$$

Then

$$\Gamma_{R}{}^{\sigma}{}_{\lambda\mu} = -\tfrac{1}{2}i\{\textstyle\sum_k A_{Rk}(x'')_{\mu}[\tau_k]^{\sigma}{}_{\lambda} + \sum_k A_{Rk}(x'')_{\lambda}[\tau_k]^{\sigma}{}_{\mu}\} \quad (50.14)$$

$$= A_{R}{}^{\sigma}{}_{\mu\lambda} + A_{R}{}^{\sigma}{}_{\lambda\mu}$$

(summed over k) with the matrix $A_{R}{}^{\sigma}{}_{\mu\lambda}$ given by

$$A_{R}{}^{\sigma}{}_{\mu\lambda} = -\tfrac{1}{2}i\textstyle\sum_k A_{Rk\mu}[\tau_k]^{\sigma}{}_{\lambda} \quad (50.15)$$

with $A_{R}{}^{\sigma}{}_{\mu\lambda}$ transformable to matrix row and column numbers

$$A_{R\text{flat}}{}^{\mu a}{}_{b} = A_{R\text{flat}k}{}^{\mu}[\tau_k]^{\sigma}{}_{\lambda}\delta_{\sigma}{}^{a}\delta^{\lambda}{}_{b}$$

using the flat space-time vierbein values, and so $A_{R\text{flat}}{}^{a}{}_{\mu b}$ may be written in matrix form as

$$A_{R\text{flat}\mu} = -\tfrac{1}{2}i\textstyle\sum_k A_{R\text{flat}k\mu}\tau_k \quad (50.16)$$

In the flat space-time limit the $A_{Rk}{}^{\lambda}(y)$ becomes the Coordinate Species group U(4) gauge fields $A_{R\text{flat}k}{}^{\lambda}(y)$.

The relevant *quadratic* $A_{R}{}^{\sigma}{}_{\mu\lambda}$ terms from eq. 50.22 below that are needed to find the dynamic equation for the gauge fields $A_{R\text{flat}}{}^{i}{}_{\mu}$ are contained in

$$\mathcal{L}_A = \text{Tr}\,\sqrt{g}[M\partial_{\nu}R^{1}{}_{\sigma\mu}\partial^{\nu}R^{2\sigma\mu} + aR^{1}{}_{\sigma\mu}R^{2\sigma\mu} + bg^{\sigma\mu}(R^{1}{}_{\sigma\mu} + R^{2}{}_{\sigma\mu}) + 1/4(g_{\mu\nu} + g^{2}{}_{\mu\nu})T^{\mu\nu}] \quad (50.17)$$

We can let

$$R^{i}{}_{\sigma\mu} = R^{i\beta}{}_{\sigma\beta\mu} \equiv \partial_{\mu}(A_{R}{}^{i\beta}{}_{\sigma\beta} + A_{R}{}^{i\beta}{}_{\beta\sigma}) - \partial_{\beta}(A_{R}{}^{i\beta}{}_{\sigma\mu} + A_{R}{}^{i\beta}{}_{\mu\sigma}) \quad (50.18)$$

for i = 1, 2. In the flat space-time limit we chose the Landau gauge

$$\partial_{\mu}A_{R\text{flat}}{}^{i\mu a}{}_{b} = 0 \quad (50.19)$$

As a result

$$R^{i}{}_{\sigma\mu} \equiv \partial_\mu({A_R}^{i\beta}{}_{\sigma\beta} + {A_R}^{i\beta}{}_{\beta\sigma}) \qquad (50.20)$$

Using

$$A_{R_{flat}}{}^{i\mu\sigma}{}_{\lambda} = A_{R_{flat}}{}^{i\mu}[\tau_k]^{a}{}_{b}\delta^{\sigma}{}_{a}\delta_{\lambda}{}^{b} \qquad (50.21)$$
$$A_{R_{flat}}{}^{i}{}_{\mu} = -\tfrac{1}{2}i\sum_k A_{R_{flat}}{}^{i}{}_{k\mu}$$

and taking the trace in eq. 50.17 we obtain

$$\mathfrak{L}_A = \mathrm{Tr}\sqrt{g}[8M\partial_\nu\partial_\mu A_{R_{flat}}{}^{1}{}_{\sigma}\partial^\nu\partial^\mu A_{R_{flat}}{}^{2\sigma} + 8a\partial_\mu A_{R_{flat}}{}^{1}{}_{\sigma}\partial^\mu A_{R_{flat}}{}^{2\sigma} + 1/4(g_{\mu\nu} + g^2{}_{\mu\nu})T^{\mu\nu}] \qquad (50.22)$$

in the flat space-time limit. Eq. 50.17 needs to take account of the complex nature of $(g_{\mu\nu} + g^2{}_{\mu\nu})$ until transformed by the infinitesimal form of the complex Reality transformation:

$$\begin{aligned}(g_{\beta\alpha} + g^2{}_{\beta\alpha})' &\rightarrow U(x'')_\beta{}^\mu(g_{\mu\nu} + g^2{}_{\mu\nu})U^{-1}(x'')^\nu{}_\alpha \\ &= (\delta_\beta{}^\mu + i\sum_k\Phi_k(x'')[\tau_k]_\beta{}^\mu)(g_{\mu\nu} + g^2{}_{\mu\nu})(\delta^\nu{}_\alpha - i\sum_k\Phi_k(x'')[\tau_k]^\nu{}_\alpha) \\ &\cong (g_{\beta\alpha} + g^2{}_{\beta\alpha}) + i\{\sum_k\Phi_k(x'')[\tau_k]_\beta{}^\mu - i\sum_k\Phi_k(x'')[\tau_k]^\nu{}_\alpha)(g_{\mu\nu} + g^2{}_{\mu\nu}) \\ &\cong (g_{\beta\alpha} + g^2{}_{\beta\alpha}) + i\sum_k\Phi_k(x'')\{[\tau_k]_{\beta\alpha} - [\tau_k]_{\beta\alpha}\}\end{aligned} \qquad (50.23)$$

Approximating $\Phi_k(x'')$ with an infinitesimal line we find

$$\sum_k\Phi_k(x) \cong \delta x_\lambda A_{Rflat}{}^{1\lambda}(x) \qquad (50.24)$$

by eq. 50.13. Thus

$$\tfrac{1}{4}(g_{\mu\nu} + g^2{}_{\mu\nu})T^{\mu\nu} \cong \tfrac{1}{4}[(g_{\beta\alpha} + g^2{}_{\beta\alpha}) + i\,\delta x_\lambda A_{Rflat}{}^{1\lambda}(x)\{[\tau_k]_{\beta\alpha} - [\tau_k]_{\beta\alpha}\}]T^{\alpha\beta}$$

Applying the canonical Euler-Lagrange method we obtain the dynamical equations (using integration by parts to handle higher order derivative terms):[82]

$$\Box^2 A_{R_{flat}}{}^{1}{}_{\sigma} + (a/M)\Box A_{R_{flat}}{}^{1}{}_{\sigma} + i\delta x_\sigma\{[\tau_k]_{\beta\alpha} - [\tau_k]_{\beta\alpha}\}T^{\alpha\beta}/(32M) = 0 \qquad (50.25)$$
$$\Box^2 A_{R_{flat}}{}^{2}{}_{\sigma} + (a/M)\Box A_{R_{flat}}{}^{2}{}_{\sigma} = 0 \qquad (50.26)$$

Since the $A_{R_{flat}}$ gauge field is gravitational in nature it exists, as eq. 50.25 shows, as a type of gravitational interaction whose source is the energy-momentum tensor. Following the standard derivation of the gravitational potential we find the Coulomb interaction of $A_{R_{flat}}{}^{10}$.

---

[82] It is possible that the Reality transformation also depends on $A_{R_{flat}}{}^{2}{}_{\sigma}$. Then eq. 50.26 would have an energy-momentum tensor term as well. Consequently there would be an additional interaction of the same form as in eq. 50.27 below.

51. Species Group U(4) Gauge Fields

From the discussion in sections 50.4 - 50.5 we see the flat space-time limit of $A_{Rk}{}^{\lambda}(y)$ is a local U(4) coordinate space gauge field. There are, *by assumption,*[83] a corresponding internal symmetry gauge fields $A_{Sk}{}^{\lambda}(y)$ – the Internal Symmetry U(4) Species Group gauge fields. The mathematical features of this field is quite similar to the U(4) Generation group fields. The interaction that appears in covariant derivatives is $g_8\mathbf{A}_S{}^{\mu}(x) = g_8\mathbf{A}_{Sk}{}^{\mu}(x)\mathbf{G}_{Sk}$ where the $\mathbf{G}_{Sk}$ are U(4) generator matrices and k is summed from 1, … , 16.

Below we will see that the effect of the Internal Symmetry Species Group is to U(4) rotate the four components of each fermion's field. Since it preserves the species of each fermion we call this group the *Species Group*. It performs a U(4) rotation of the spinor representation of each fermion

We will see that the Higgs Mechanism breakdown of the Species Group endows each fermion with a mass contribution that breaks the scale invariance of the Unified SuperStandard Theory.

Since the Species Group Higgs Mechanism breaking gives each fermion a 'gravity generated' mass, and since this mass sets the mass scale for each fermion, we conclude later that the principle of the equality of inertial and gravitational mass is a direct consequence. ***Inertial mass equals gravitational mass.***

51.1 Species Group Covariance

A Species Group transformation on a Dirac equation must be covariant. Consider the Dirac equation lagrangian term under an Internal Symmetry Species Group transformation:

$$\bar{\psi}(x)[i\gamma_\mu(\partial/\partial x_\mu - ig_8A_{Sk}{}^{\mu}(x)G_{Sk}) - m]\psi(x) = 0 \tag{51.1}$$

summed over k. If we perform a U(4) Species group transformation U on lagrangian terms:

$$\bar{\psi}(x)[i\gamma_\mu(\partial/\partial x_\mu - ig_8A_{Sk}{}^{\mu}(x)G_{Sk}) - m]U^{-1}U\psi(x)$$

or

$$\bar{\psi}(x)U^{-1}U[iU^{-1}U\gamma_\mu U^{-1}U(\partial/\partial x_\mu - ig_8A_{Sk}{}^{\mu}(x)G_{Sk}) - m]U^{-1}U\psi(x)$$

we find

$$\bar{\psi}'(x)[i\gamma_\mu'U(\partial/\partial x_\mu - ig_8A_{Sk}{}^{\mu}(x)G_{Sk})U^{-1} - m]\psi'(x)$$

where

$$\gamma_\mu'(x) = U\gamma_\mu U^{-1}$$

---

83 In this discussion we *assume* that the Coordinate Species Reality Group with gauge fields $A_{Rk}{}^{\lambda}(y)$ has a corresponding Internal Symmetry Reality group that we call the Internal Symmetry Species Group. This assumption parallels the assumptions for the SU(3)⊗SU(2)⊗U(1)⊗SU(2)⊗U(1) Internal Symmetry Reality Group presented in previous chapters.

is locally equivalent to a Dirac matrix by Good's Theorem.[84] If we set

$$A'_S{}^{\mu}(x) = -(i/g_8)U[\partial U^{-1}/\partial x^{\mu}] + UA_S{}^{\mu}(x)U^{-1}$$

then the transformed lagrangian terms are

$$\bar{\psi}'(x)[i\gamma_{\mu}'(x)(\partial/\partial x_{\mu} - ig_8A'_{Sk}{}^{\mu}(x)G_{Sk}) - m]\psi'(x) \tag{51.2}$$

They have the same form as the original terms above and thus the expression is covariant. We note the indices of the matrices G_{Sk} are spinor indices and so $G_{Sk}\gamma_{\mu}$ has an implicit spinor matrix summation. But the symmetry group is U(4).

The coordinate dependence of $\gamma_{\mu}'(x)$ introduces locality into the Dirac matrix. This locality might be viewed with concern except that an inverse Species group transformation exists that removes the locality. Thus the physical impact of this 'new' locality is eliminated.

51.2 Spontaneous Symmetry Breaking of the General Relativity U(4) Reality Group – The Species Group

We begin the discussion of the Internal Symmetry Species Group symmetry breaking[85] by defining a Higgs field η which is a Species group 4-vector

$$\eta = \begin{bmatrix} \rho_1 \\ \rho_2 \\ \rho_3 \\ \rho_4 \end{bmatrix} \tag{51.3}$$

where ρ_1, ρ_2, ρ_3 and ρ_4 are real fields.[86] Then the covariant derivative of η (taking account only of the Species group) is

$$D_{\ldots\mu}\eta = \{\partial/\partial X^{\mu} + \ldots - \tfrac{1}{2}ig_8\Sigma A_{Sk}{}^{\mu}(x)G_{Sk}\} \begin{bmatrix} \rho_1 \\ \rho_2 \\ \rho_3 \\ \rho_4 \end{bmatrix} \tag{51.3}$$

---

[84] R. H. Good, Jr., Rev. Mod. Phys., **27**, 187 (1955).

[85] Since the Species gauge fields have been shown to have a mass it might seem redundant to introduce Higgs symmetry breaking as well. However the Higgs breaking introduces the further benefit of giving a mass term to each particle – thus establishing the equality of gravitational mass and inertial mass as we discuss in section 20.4.

[86] Each field ρ_i can be expressed as a PseudoQuantum field: $\rho_i = \varphi_{1i} + \varphi_{2i}$ where φ_{1i} has the vacuum expectation value $\rho_{i0.\ \text{for } i = 1, \ldots, 4.}$ Thus our PseudoQuantum field theory version is implemented easily.

Following steps similar to eqs. 47.4 through 47.17 for the Generation Group symmetry breaking we find with ρ_i being the vacuum expectation value of the Higgs field:

$$\begin{aligned}(D_{\dots\mu}\eta)^\dagger D_{\dots}^{\ \mu}\eta &= \partial\rho_1/\partial X^\mu\, \partial\rho_1/\partial X_\mu + \partial\rho_2/\partial X^\mu\, \partial\rho_2/\partial X_\mu + \partial\rho_3/\partial X^\mu\, \partial\rho_3/\partial X_\mu + \\ &+ \partial\rho_4/\partial X^\mu\, \partial\rho_4/\partial X_\mu + \\ &+ \tfrac{1}{4} g_8^2\{\rho_1^2 A_{S1}^2 + \rho_2^2 A_{S2}^2 + \rho_3^2 A_{S3}^2 + \rho_4^2 A_{S4}^2 + \\ &+ (\rho_1^2 + \rho_2^2)(V_5^2 + V_6^2) + \tfrac{1}{4}(\rho_1^2 + \rho_3^2)(V_7^2 + V_8^2) + \\ &+ (\rho_1^2 + \rho_4^2)(V_9^2 + V_{10}^2) + \tfrac{1}{4}(\rho_2^2 + \rho_3^2)(V_{11}^2 + V_{12}^2) + \\ &+ (\rho_2^2 + \rho_4^2)(V_{13}^2 + V_{14}^2) + \tfrac{1}{4}(\rho_3^2 + \rho_4^2)(V_{15}^2 + V_{16}^2)\}\end{aligned} \tag{51.4}$$

up to total divergences, which generate surface terms which we discard. We also assume that all fields satisfy the gauge condition

$$\partial A_{Si}^{\ \mu}/\partial X^\mu = 0 \tag{51.5}$$

Eq. 51.4 shows all Species Group gauge fields have masses. Thus Species Group symmetry is completely broken. The combination of an ultra-weak coupling constant and very large gauge field masses results in extremely Species interactions.

We assume Species group gauge field masses to be very large – of the order of the Planck mass in view of its origin in Complex General Relativity.

51.3 Species Group Higgs Mechanism Contributions to Fermion Masses

The symmetry breaking of the Species Group results in a contribution to each fermion mass of all types, species, generations, and layers. The Species Group contributions to normal and Dark fermion mass terms are

$$\mathscr{L}^{Higgs}_{FermionMassesSpecies} = \Sigma_{s,g,l}\bar{\psi}_{sglL}\rho_s\, m_{sgl}\psi_{sglR} + \Sigma_{s,g,l}\bar{\psi}_{DsglL}\rho_s\, m_{Dsgl}\psi_{DsglR} + \text{c.c.} \tag{51.6}$$

The η field expectation value has components labeled ρ_s.[87] The mass matrices m_{sgl} and m_{Dsgl} are the complex constant Species mass matrix contributions for normal and Dark species.

51.4 Species Group Higgs Masses Shows Inertial Mass Equals Gravitational Mass

In Blaha (2016h) we showed that a Complex General Relativity transformation can be factored into the product of a complex-valued transformation and a real-valued General Coordinate transformation. The set of complex valued transformations form a U(4) group that we called the General Coordinate Reality group. The analogous Internal Symmetry Species Group has gauge fields that undergo spontaneous symmetry breaking and generate contributions to all fermion masses.

[87] The Higgs fields $\eta_{\dots}$ in our PseudoQuantum formulation are $\eta_{\dots} = \varphi_{1\dots}(x) + \varphi_{2\dots}(x)$ as described earlier.

Since fermion field masses are now sums of ElectroWeak Higgs contributions, Generation group Higgs contributions, Layer group Higgs contributions, and General Coordinate Species Group contributions, and since the Species Group Higgs fields appear in all fermion masses, the equality of inertial and gravitational mass is proven. The Species Group Higgs particles' equations set the mass scale of gravitational mass, and thereby of all Higgs mass contributions. The scale of inertial masses equals to the scale of gravitational masses. ***Since an expression cannot mix mass scales, the gravitational mass scale must be the same as the inertial mass scale.***

Inertial mass equals gravitational mass.

We have established the equality of inertial and gravitational mass at the short distance quantum level. In our view, this explanation is far more satisfying than basing the equality on a combination of large distance phenomena and quantum phenomena. As Einstein and Weyl have pointed out, all fundamental physics phenomena should be based on a local theory.

We have mapped Complex General Relativistic transformations consisting of U(4) transformations and Real General Relativistic transformations into transformations consisting of Internal Symmetry Species Group factors and Real General Relativistic transformations factors. The Higgs Mechanism breakdown of the Species Group has the important consequence that it prevents Species Group transformations that rotate between fermions and anti-fermions.

Appendix C. Some Features of QUeST Quaternion Space

This appendix describes 32 complex quaternion dimension space and some of the features of QUeST. It consists of material from Blaha (2020d) – (2020i).

C.1 Thirty-Two Complex Quaternion Space – 32 × 8 Dimension Array

A quaternion contains four dimensions. A complex quaternion contains eight dimensions. It is a complexification of the quaternion concept. Fig. C.1 depicts the 32 dimension space. It uses a "dot" • to represent a dimension. The dimensions of the space are not assigned physically until they are mapped to internal symmetry group fundamental representation dimensions and space-time dimensions. Rather than create a cumbersome coordinate-based notation we choose to use •'s.

Figure C.1. The 32 complex quaternion dimension QUeST array. This array is the 32 × 8 array of •'s.

The 32 × 8 form of the array is useful because it brings out the four layers of fermions that appear when the array is subdivided into four layers (8 rows each) of fundamental group representations. The Unified SuperStandard Theory[88] implied by QUeST has a matching four layers of fermions. The subdivision of Fig. C.1 into layers appears in Fig. C.2. The map to group representations appears in Figs. C.3 and C.4. We use the maps in Table 1.1 to set up the group ↔ dimension map, bearing in mind the group representations of the Standard Model:

[88] Blaha (2020d) and earlier books.

U(1)	↔	2 real dimensions
U(4)	↔	8 real dimensions
U(2)	↔	4 real dimensions
SU(3)	↔	6 real dimensions
U(1)⊗SU(2)	↔	4 real dimensions

Table C.1. Map between fundamental representations and their dimensions.

where the dimensions have *real-valued* coordinates and are called *real dimensions*.

The eight U(4) Layer groups and eight U(4) Generation groups in Fig. C.3 are present in both UST and QUeST as well as being implied by BQUeST. See Fig. C.4 for the groups of one layer of the four layers.

C.1.1 QUeST Internal Symmetry Groups

The QUeST internal symmetry group is

$$[SU(2)\otimes U(1)\otimes SU(3)]^8\otimes U(4)^{16}\otimes U(2)^4 \qquad (C.1)$$

by Fig. C.3 below.

C..1.2 QUeST Space-Time

The space-time of QUeST is 4 complex quaternion dimensions.

C.1.3 Generation and Layer groups of UST, QUeST and BQUeST

The Generation groups mix the fermion generations of normal and Dark sectors of each layer. The lines on the left side of Fig. C.5 display Generation group mixing within each layer.

Layer groups mix fermions in all four layers for each of the four generations individually. (See right side of Fig. C.5.) There are eight Layer groups: two Layer groups for Normal and Dark sectors for each generation.

The Dark groups mix between normal and Dark fermions, fermion by fermion.

C.1.4 Fermion- Dimension Duality

Fig. C.6 shows a 1:1 relation between QUeST dimensions and the fundamental fermions of QUeST and UST. This duality is the basis of the BQUeST and BMOST one-dimension theories implying QUeST and UTMOST respectively.

C.1.5 Fermion Structure Extracted from QUeST Symmetry Structure

Given the form of the internal symmetries in QUeST we can determine the fermions in the fundamental group representations as shown in Fig. C.5.

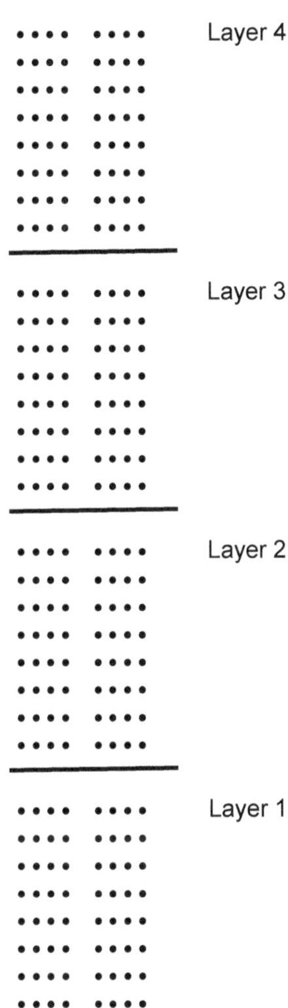

Figure C.2. The 32 complex quaternion dimension QUeST array subdivided into 4 layers of 8 rows..Each layer will be seen to map to a block of fundamental group representations as shown in Figs. C.3 and C.4.

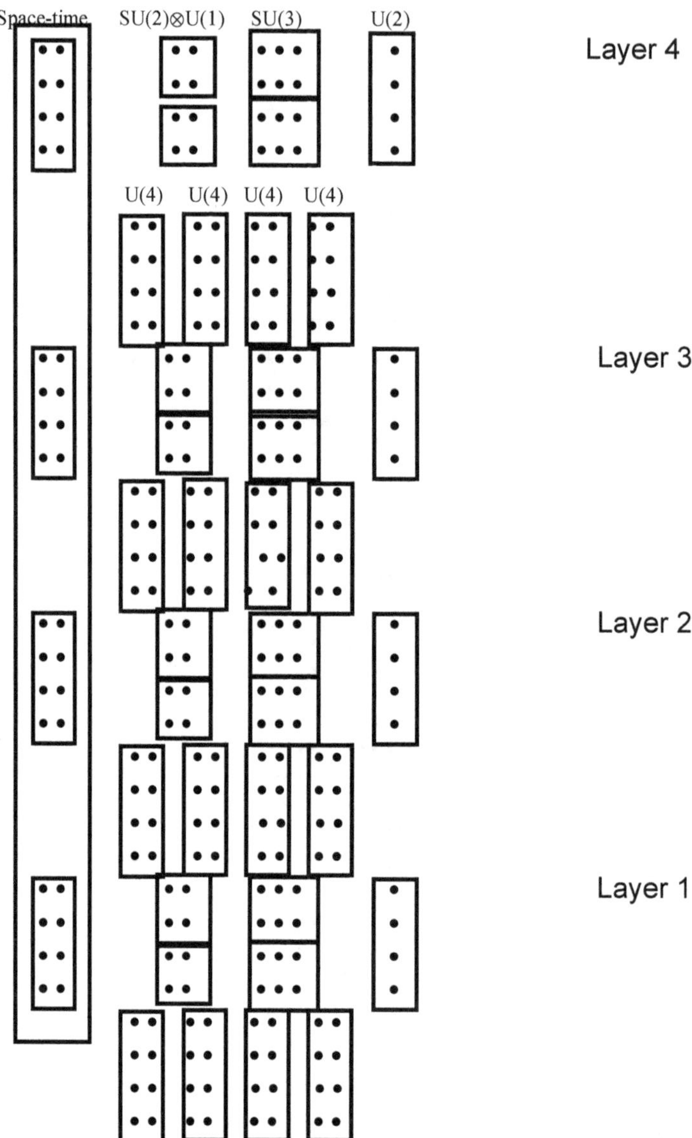

Figure C.3. The four layers of QUeST internal symmetry groups (and space-time) for 32 dimension complex quaternion space. Note: each row has an 8 • complex quaternion. Note the left column of blocks combine to specify a 4 dimension complex quaternion space-time. Note each layer requires 64 dimensions. *The U(2) Dark group is broken to U(1)⊗U(1) in chapter 2 to conform to the breakdown to 4 × 4 blocks suggested by comparison to the spinor structure.*

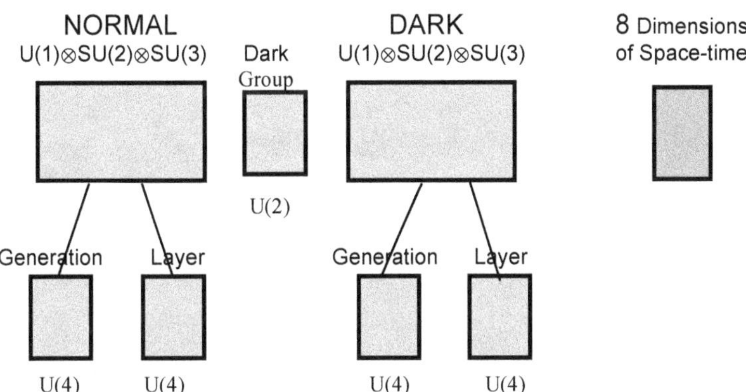

Figure C.4. The internal symmetry groups for one QUeST layer of the 4 layers in the 32 × 8 dimension array format. The two large blocks are each 5 complex dimension (10 real dimension) representations of SU(2)⊗U(1)⊗SU(3). The U(2) group (badly broken) supports transformations (rotations) between Normal and Dark matter.

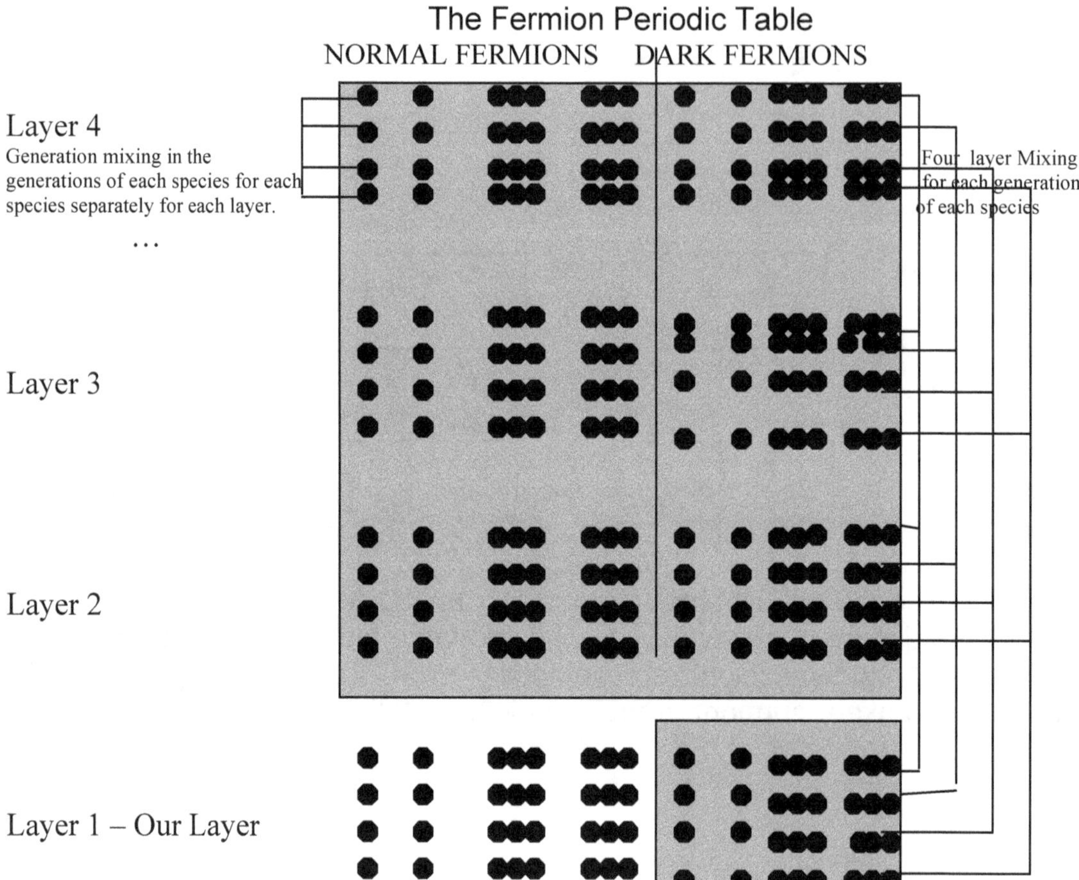

Figure C.5. Fermion particle spectrum and partial examples of the pattern of mass mixing of the Generation group and of the Layer group. Unshaded parts are the known fermions with an additional, as yet not found, 4th generation. The lines on the left side (only shown for one layer) display the Generation mixing within each layer. The Generation mixing occurs within each layer using a separate Generation group for each layer. The lines on the right side show Layer group mixing (for Dark matter) with the mixing among all four layers for each of the four generations individually. There are four Layer groups for Normal matter and four Layer groups for Dark matter.. There are 256 fundamental fermions. QUeST and UST have the same fermion spectrum.

	QUATERNION DIMENSIONS		FERMIONS			
	Real	Imaginary	e	v	up-q	down-q
Layer 1	••••	••••	•	•	•••	•••
	••••	••••	•	•	•••	•••
	••••	••••	•	•	•••	•••
	••••	••••	•	•	•••	•••
			DARK			
			e	v	up q	down q
	••••	••••	•	•	•••	•••
	••••	••••	•	•	•••	•••
	••••	••••	•	•	•••	•••
	••••	••••	•	•	•••	•••
Layer 2	••••	••••	•	•	•••	•••
	••••	••••	•	•	•••	•••
	••••	••••	•	•	•••	•••
	••••	••••	•	•	•••	•••
	••••	••••	•	•	•••	•••
	••••	••••	•	•	•••	•••
	••••	••••	•	•	•••	•••
	••••	••••	•	•	•••	•••
Layer 3	••••	••••	•	•	•••	•••
	••••	••••	•	•	•••	•••
	••••	••••	•	•	•••	•••
	••••	••••	•	•	•••	•••
	••••	••••	•	•	•••	•••
	••••	••••	•	•	•••	•••
	••••	••••	•	•	•••	•••
	••••	••••	•	•	•••	•••
Layer 4	••••	••••	•	•	•••	•••
	••••	••••	•	•	•••	•••
	••••	••••	•	•	•••	•••
	••••	••••	•	•	•••	•••
	••••	••••	•	•	•••	•••
	••••	••••	•	•	•••	•••
	••••	••••	•	•	•••	•••
	••••	••••	•	•	•••	•••

Figure C.6. Fundamental fermions have a 1:1 correspondence with QUeST dimensions. Note the number of dimensions in each row is 8 – the number of dimensions in a complex quaternion. Correspondingly the number of fermions in each row is 8 – a suggestive similarity. Each layer has four normal fermion generations and four Dark fermion generations. Each dot (pebble) represents a dimension in the left part of the figure and a fermion in the right part.

C.2 Thirty-Two Complex Quaternion Space – 16 × 16 Dimension Array

This section describes an alternate 16 × 16 form of the QUeST dimension array. This format supports the derivation of the QUeST dimensions from BQUeST.

The 16 × 16 form of the QUeST dimension array can be based on a 16 complex octonion dimension space. The difference between this format space and the 32 dimension space is not physically meaningful at present. The difference will be physically meaningful if the masses of the fermion spectrum and the full pattern of symmetry breaking is determined. Then one can differentiate between a four layer theory as above in section C.1 and a two "layer" theory presented below in this section.

The alternate 16 × 16 form of the QUeST dimension array is simply constructed from the preceding 32 × 8 dimension array by "moving" the 16 × 8 lower half of the 32 × 8 array of Fig. C.1 to the "right" of its upper half. The 16 × 16 dimension array appears in Fig. C.7. Note that the forms are physically equivalent if done prior to mapping dimensions to group representations.

The 16 × 16 form of the dimension array is more convenient for our derivation of a fundamental basis for QUeST that we call BQUeST (pronounced bee-quest). No change in the number of dimensions is made. (QUeST could also be viewed as residing in a 16 complex octonion dimension space for the purpose of dimension counting.)

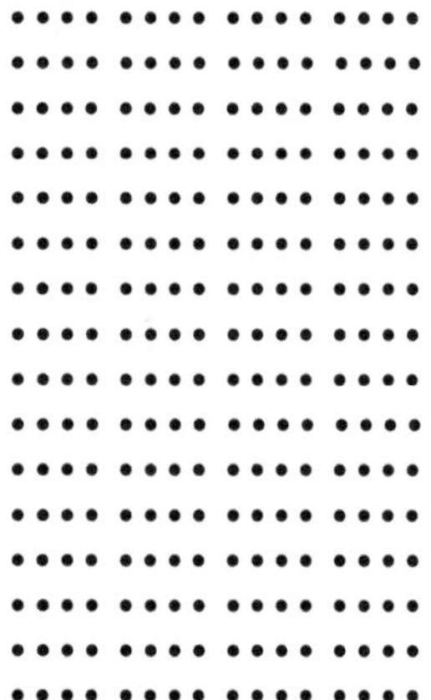

Figure C.7. The 16 × 16 array of QUeST dimensions.

Fig. C.8 shows the assignment of fundamental group representations to the 16 × 16 dimension array. Note this form of array has two "layers" of 128 dimensions in contrast to the four 64 dimension layers of Fig. C.3. The difference can be attributed to the third and fourth layers of Fig. C.3 becoming "extra" Dark sectors in "layers" 1 and 2.

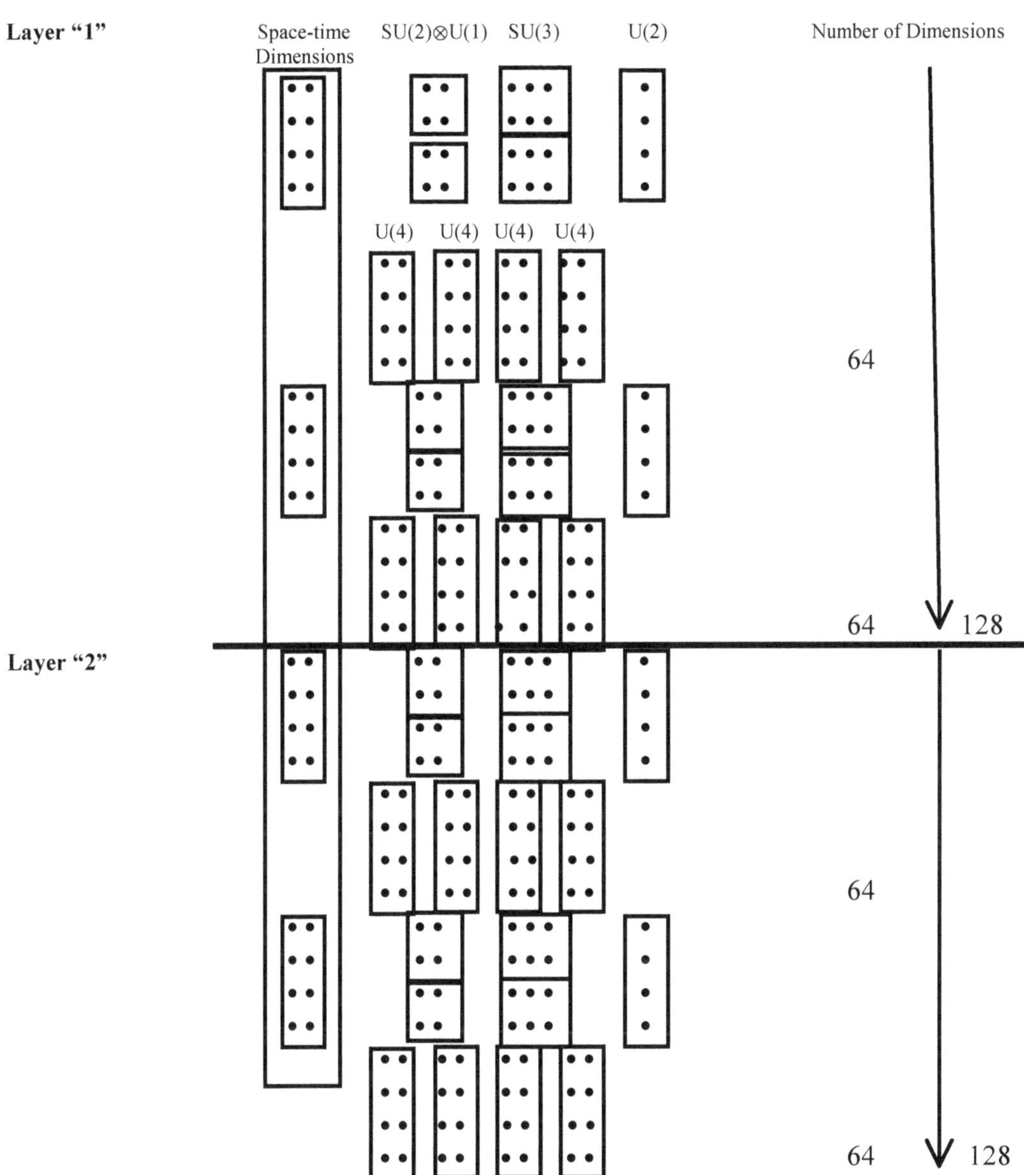

Figure C.8 The 16 × 16 form of QUeST array has two "layers", each of which is composed of two layers of the 32 × 16 four layer figure of Fig. C.3. Note each of these "layers" has 128 dimensions.

The group representations in Fig. C.8 can be subdivided into blocks of 16 dimensions as shown in Fig. C.9 and C.10 for the "layers" of Fig. C.8. Blaha (2020i) provides a rationale for 16 dimension subblocks based on U(8) considerations.

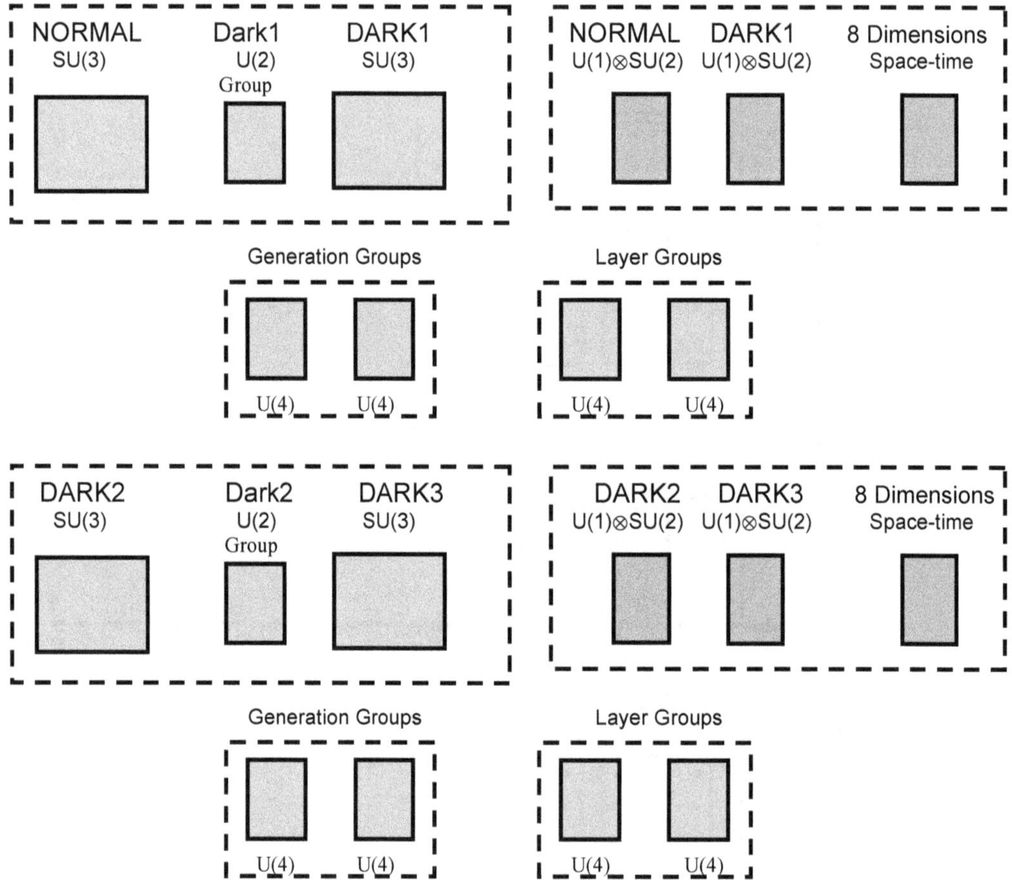

Figure C.9. Set of 16 dimension blocks for the first "layer" (of two "layers") of the 16 × 16 array. Each "dashed" block (regardless of its apparent size) is a 4 × 4 = 16 array of dimensions. This set of 8 blocks contains the 8×16 = 128 dimensions of "layer" 1. "Layer" 2 is similar. The Dark U(2) groups supports transformations (rotations) between the types of matter: Normal and Dark1; and Dark2 and Dark3.

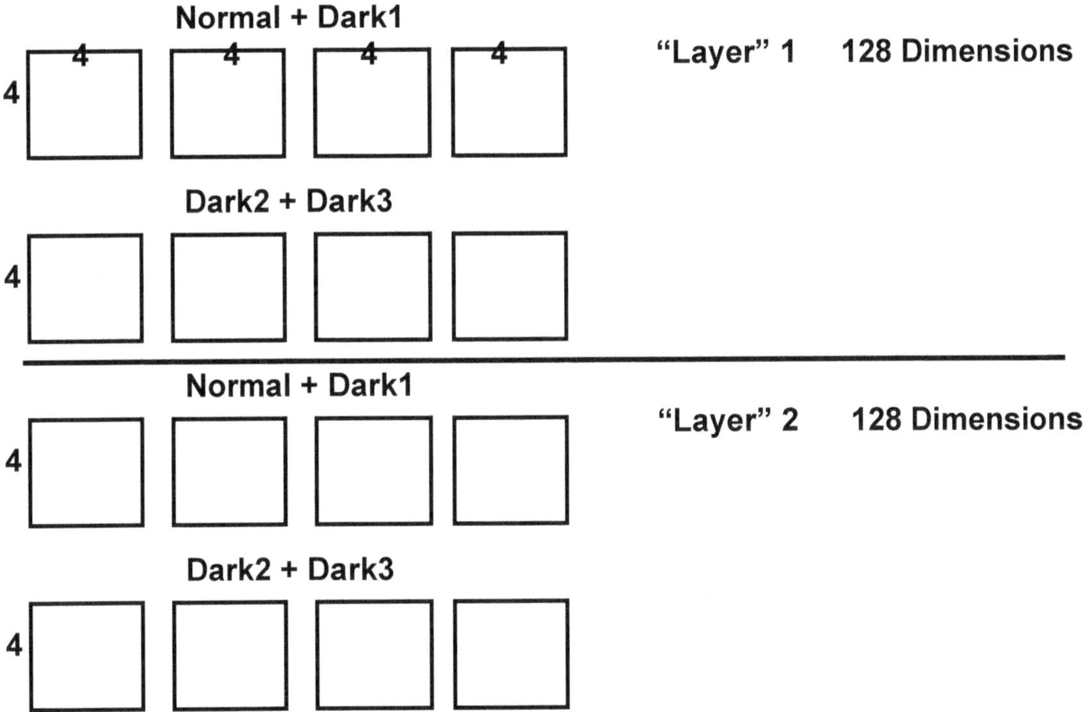

Figure C.10. The two "layers" of 4 × 4 dimension subblocks of the 16 × 16 dimension array. .

QUeST fermions have a similar format to the dimension subblock structure in Fig. C.10. Fig. C.11 displays the 256 fermions of QUeST. The shift from a 32 × 8 dimension array with four layers of groups to a 16 × 16 dimension array gives a two "layer" form with the" lower" two layers of the 32 × 8 dimension array becoming the Dark2 and Dark3 parts of the two "layer" form. This shift has no discernable physical consequences as far as our considerations are concerned since we are moving Dark sectors. When mass generation and symmetry breaking are better understood, a physically significant difference may be evident.

Normal Dark1 Dark2 Dark3

Figure C.11. Spectrum of the generations of fermions of QUeST for the 16 × 16 dimension array representation. Each fermion is represented by a •. Quark triplets are represented by three •'s. Note there are 256 fundamental fermions.

The possible U(8) 4 × 4 block structure of Figs. C.9 – C.10 suggest a similar block structure for the fundamental fermions. Fig 1-A.12 displays a set of sixteen 4 × 4 blocks with each block holding 16 fermions. It has a SU(4)-like structure of the four rows (generations) of each 16 dimension subblock. The subblocks have a lepton-triquark content. Clearly the implied SU(4) symmetry would be broken. It is suggestive of a Lorentz 4-vector representation with the lepton corresponding to a time coordinate and the three quarks corresponding to spatial coordinates.

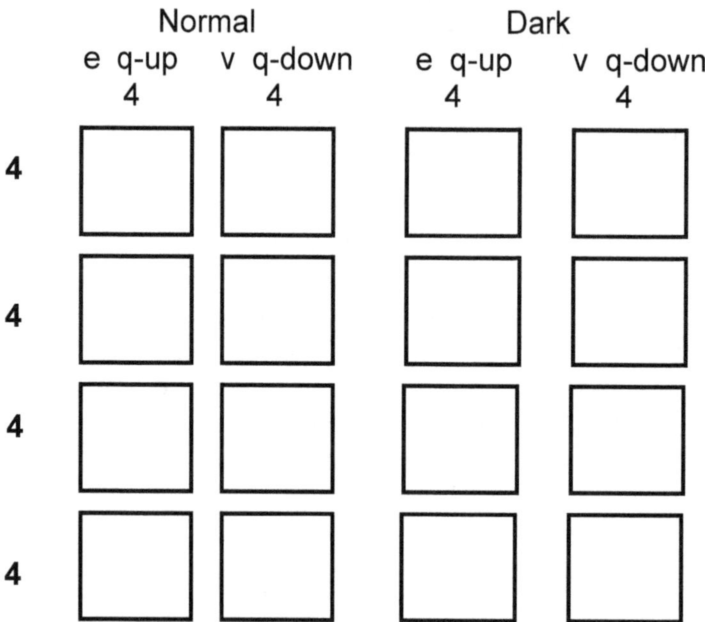

Figure C.12. Block form of a 16 × 16 QUeST fermion array with each block row corresponding to one layer. Each block contains four generations of fermions. The result is 4 × 4 blocks. The label e q-up indicates a charged lepton – up-type quark pair, ν q-down indicates a neutral lepton – down-type quark pair, and so on. Note the blocks can be reaaranged into a 32 × 8 form without physical consequences at this level of discussion since the right two columns and the lowest two rows are all Dark at present.

C.3 A Partition to Real 3+1 Dimension Space-times

We can partition Fig. C.4 into real and imaginary subspaces with 3+1 dimension space-times.

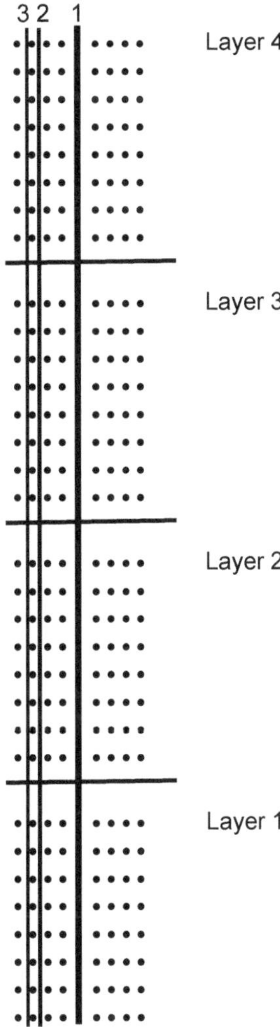

Figure C.13. The 32 complex quaternion dimension QUeST array partitioned down to a real spwce-time. The partition labeled “1” reduces the array to 32 quaternion dimensions after discarding the right columns. The partition labeled “2” reduces the array to 32 complex dimensions similarly. The partition labeled “3” reduces the array to 32 real-valued dimensions in the leftmost column..

The real-valued column of dimensions numbers 32 dimensions. It maps to the internal symmetry group:

$$[SU(2)\otimes U(1)\otimes SU(3)]^2\otimes U(4) \tag{C.2}$$

where the U(4) group is the Generation group, which implies 4 generations. There is an SU(2)⊗U(1)⊗SU(3) group giving the ElectroWeak and Strong interactions of Normal matter. The other SU(2)⊗U(1)⊗SU(3) group gives the Dark Matter ElectroWeak and Strong Interactions. Note the Layer group is not present.

These groups require 28 dimensions. The remaining four dimensions give 4 dimension space-time. The total is 32 dimensions.

The fundamental fermions corresponding to the internal symmetry groups number 32 fermions (fermion-dimension duality). See Fig. C.14.

Normal Dark
•••••••• ••••••••
•••••••• ••••••••
•••••••• ••••••••
•••••••• ••••••••

Figure C.14. Spectrum of the generation of fermions. Note only one layer of fermions. Two leptons and 6 quarks in each row of Normal matter. Similarly for Dark matter.

This restricted model is similar to The Standard Model: same internal symmetry, same 4 generations, same space-time, and same fermion and vector boson spectrums..

Appendix D. Evidence for Universe Particles

This appendix provides evidence for the existence of a Megaverse and for the possibility that universes are particles. It is abstracted from Blaha (2018e), which provides much more detailed data.

D.1 Evidence for Entities Beyond the Universe

At first glance it would seem impossible to produce evidence for the existence of other universes. However there are subtle means by which we can 'sense' experimentally 'nearby' universes should they exist. The mechanism would appear to be gravitational effects exerted on objects within our universe by unseen objects of enormous mass. Currently there appears to be three experimental suggestions of the existence of 'nearby' universes and one theoretical argument based on an influx of mass-energy from the Megaverse that may cause the expansion of our universe.

There are also theoretical motivations for believing that there are entities beyond our universe. These are detailed in chapter 30 of Blaha (2018e).

D.1.1 Great Attractors

One potential support is the discovery of the Great Attractor (at the center of the Laniakea Galaxy Supercluster), and the more massive Shapley Attractor (centered in the Shapley Supercluster)[89]. These attractors contain massive numbers of galaxies and are drawing galaxies over a distance of millions of light years towards them.

If another universe(s) is 'near' our universe it could act as a 'gravitational magnet' and draw galaxies within our universe towards it to form one or more superclusters which could then act as attractors. Thus attractors might indirectly reveal the presence of other nearby universes—contrary to the expected large scale uniformity of the universe. The only other apparent source of superclusters is chance. Chance seems an unsatisfactory possibility in the present case.

D.1.2 Bright Bumps in Universe Suggesting Collision with Another Universe

A recent study[90] of the residual brightness of parts of the accessible universe found that bright patches appeared if a model of the CMB (Cosmic Microwave Background) with gases, stars and dust was 'subtracted' from the PLANCK map of the entire sky. After the subtraction one would expect only noise spread throughout the sky. However, bright patches were seen in a certain range of frequencies. These anomalies are thought to be a result of our universe colliding with another object – presumably another universe in the Megaverse.

---

[89] Tully, R. Brent; Courtois, Helene; Hoffman, Yehuda; Pomarède, Daniel, "The Laniakea Supercluster of galaxies". Nature (4 September 2014). 513 (7516): 71–73; arXiv:1409.0880.

[90] Ranga-Ram Chary, arXiv.org:/1510.00126 (2015).

D.1.3 Cold Spot in Universe Suggesting Collision with Another Universe

Another recent study[91] of a huge cold region of the universe spanning billions of light years revealed that this region is not a relatively empty region but rather is similar to in its distribution of galaxies to the rest of the universe. Previous the Cold Spot (an area where cosmic microwave background radiation – the leftover Big Bang radiation is weak – making it significantly colder (0.00015C colder) than the average temperature of the universe.)

An analysis of 7,000 galaxy redshifts using new high-resolution data has now shown that the Cold Spot is similar to the rest of the universe. The Durham University group suggested that the Cold Spot might have been caused by a collision between our universe and another Universe. They further suggested that there is only a 1 in 50 chance that it could be explained by standard cosmology.

Thus we have another important piece of circumstantial evidence in favor of other universes and thus the Megaverse.

D.1.4 Megaverse Energy-Matter Infusion into Our Universe

In chapter 14 of Blaha (2017c) we presented a model for an influx of mass-energy from the Megaverse to support the Bondi-Gold-Hoyle-Narlikar Steady State Cosmology, which was originally based on the 'continuous creation of mass-energy' by Hoyle and Narliker. This model explains why the value of Ω makes the universe close to flat. If this model is correct then we would have concrete support for a Megaverse with a low mass-energy density leaking mass-energy into our universe. *More generally, it suggests that universes are surfaces of high mass-energy density in a Megaverse of low mass-energy density – with a ratio of mass-energy densities of the other of 10^{30}.*

D.1.5 Conclusion

We conclude that data is beginning to emerge favoring multiple universes and a physical Megaverse in support of the theoretical justifications presented earlier.[92]

D.2 Hubble Constant and Universe Expansion

Our universe is clearly expanding from an initial state called the Big Bang to its current state. The Hubble "Constant" measures the rate of growth. Much of the expansion data on the Hubble Constant is presented below. After reviewing the data we propose a fit to the data in this chapter that explains the apparent growing rate of expansion. Below we derive the form of the fit.

D.2.1 Hubble Constant Experimental Data

There are a number of astrophysical studies of the universe that suggest that the Hubble Constant is *not* constant. Although there are significant margins of error it appears that the early universe "beginning" epoch around 380,000 years had a Hubble Constant of 67.8 km s^{-1} Mpc^{-1}.[93] More recently, red shift studies of quasars have given a

---

[91] T. Shanks et al, Durham University (Australia), Monthly Notices of the Royal Astronomical Society, 2016 .
[92] Chapter 59 of Blaha (2020c).
[93] See, for example, K. Aylor *et al*, arXiv:1811.00537v1 (2018) based on studies of the cosmological sound horizon.

Hubble Constant of 73.2 km s^{-1} Mpc^{-1}.[94] And studies of binary black hole merger gravity waves[95] have given a Hubble Constant of 75.2 km s^{-1} Mpc^{-1} (and earlier of 78 km s^{-1} Mpc^{-1}). Another study of events at 1.8 billion ly yielded a Hubble Constant of 70.0 km s^{-1} Mpc^{-1}.[96] Further studies have given the Hubble Constants: 1) Of variable stars 73.2 km s^{-1} Mpc^{-1}, 2) Of light bent by distant galaxies 72.5 km s^{-1} Mpc^{-1}, 3) Of Magellan Cepheids 74.03 ± 1.42 km s^{-1} Mpc^{-1}, [97] 4) Of distant red giant[98] brightness 69.8 km s^{-1} Mpc^{-1},

The only apparent conclusion at this time is that there was a Hubble Constant (Constant) H of approximately 67.8 km s^{-1} Mpc^{-1} early in the universe, and ranging up to 75.2 km s^{-1} Mpc^{-1} at the current time. Thus an increasing Hubble Constant.

For the purpose of discussing the apparent increase in H with time, we average the above eight "recent" values of H in the spirit of Bayesian equal probability to obtain a **recent time Hubble average of 73.24** km s^{-1} Mpc^{-1}.[99] Thus there appears to be a 7% - 9% increase in the Hubble Constant over time.

D.2.2 Fit to the Hubble Constant Data and Scale Factor

It is generally expected that the Hubble Constant will decline with time from the time of the Big Bang. It is generally believed that the Hubble Constant has recently been increasing with time. **The declining value in the past and the current growth of the Hubble Constant imply that it reached a minimum at some time in the past.**

Our fit to the data from Blaha (2019c) and (2019e) was

$$\begin{aligned} a(t) &= (t/t_{now})^{g + ht} \\ &= \exp[(g + ht)\ln(t/t_{now})] \end{aligned} \tag{D.1}$$

where g and h are constants. (The constant h is *not* the Hubble parameter.) There is an "ht" term in the exponent based on the rise in H(t) suggested by experimental data.

Eq. D.1 can be approximated by

$$a(t) = (H_0 t)^{g + H_0 t} \tag{D.2}$$

which nicely relates it to the Hubble parameter H_0.

The basis of the fit for a(t) was:

1. Power law behavior (in part) as in the radiation and matter dominated approximations.

---

[94] M. Soares-Santos *et al* , arXiv:1901.01540 (2019).
[95] DES and LIGO collaborations *et al*, arXiv:1901.01540 (2019).
[96] B.P. Abbott *et al*, arXiv:1710.05835 (2017).
[97] J. T. Nielsen *et al*, Marginal evidence for cosmic acceleration from Type Ia supernovae, Nature Scientific Reports (2016); arXiv.1506.01354 (2015). A. Riess *et al*, The Astrophysical Journal **875**, 145 (2019) and references therein. A. Riess *et al*, arXiv:1903.07603 (2019).
[98] W. Freedman *et al*, The Astrophysical Journal **880** (July, 2019).
[99] In Blaha (2019c) and (2019e) we used an average estimate of 73.7 km s^{-1} Mpc^{-1}.

2. The known shape of H(t) at early times, and at present, as described above

3. The simplicity of the fit. Two values of H(t) set the constants g and h.

4. Faster than exponential future growth with no Big Rip.

The Hubble Constant implied by eq. D.1 is

$$H(t) = (da/dt)/a = g/t + h(1 + \ln(t/t_{now}) \tag{D.3}$$

We set the value of H(t) by using its value at two values of time determining g and h. Based on experimental data:

$$H(t_c) \equiv H(380{,}000\text{ yr}) = 67.8\ \text{km s}^{-1}\ \text{Mpc}^{-1} \qquad \text{(D,4)}$$
$$H(t_{now}) = 73.24\ \text{km s}^{-1}\ \text{Mpc}^{-1}$$

and

$$h = (t_c H(t_c) - t_{now}H(t_{now}))[\ t_c - t_{now} + t_c \ln((t_c/t_{now})]^{-1} \tag{D.5}$$
$$g = (H(t_{now}) - h)\ t_{now}$$

where t_c = 380,000 years after the Big Bang.[100] We obtained

$$h = 2.25983 \times 10^{-18}\ \text{s}^{-1} = 1.49 \times 10^{-33}\ \text{eV} \tag{D.6}$$
$$g = 0.000282377 = 2.82377 \times 10^{-4}$$

There are two approaches to the universal scale factor fit of eq. 59.1. One approach is based on a remarkable coincidence between the power g in the fit and the QED power g seen earlier. It leads to a theory in which the expansion of the universe taken over all time is a vacuum polarization phenomenon. The other approach is based on the Einstein equation for the scale factor. We show that the Universal Scale Factor is consistent with the Einstein equation if additional (dark) energy is properly taken into account.

D.3 Hubble Parameter and Vacuum Polarization of a Particle

In this section we will show that the initial behavior of the expanding universe's scale factor is the same as the exponent of the vacuum polarization of a particle as its energy gets very large.

D.3.1 Vacuum Polarization Generation of the Early Time Part of the Universal Scale Factor

Perhaps the crowning achievement of our universal scale factor eigenvalue formulation for coupling constants is the successful relation of universe evolution to vacuum polarization due to a vector QED-like interaction between universes.

---

[100] Based on the data value of 67.8 km s^{-1} Mpc^{-1} at t = 380,000 years.

In massless QED we found that the vacuum polarization had the form:[101]

$$F_1(\alpha)(p/\Lambda)^{2g_{QED}} \qquad (D.7)$$

where $F_1(\alpha)$ is the "eigenvalue function" for the Fine Structure Constant[102] of the Johnson-Baker-Willey model of massless QED, p is the momentum, and Λ is the ultraviolet cutoff. The value of g_{QED} that corresponded to the Fine Structure Constant is

$$g_{QED} = -\ 0.00058053691948 \qquad (D.8)$$

and the Fine Structure Constant was correctly found to be

$$\alpha_{calculated}(g_{QED}) = \ 0.0072973525693 \qquad (D.9)$$

to 13 digit accuracy according to the *Particle Data Table of 2019.*

Comparing our Universal Scale Factor g value (eq. D.6) ,which governs early time behavior of the expanding universe, with g_{QED} we find

$$-g \ = 0.000282377 \cong -\tfrac{1}{2}g_{QED} = -0.000290268 \qquad (D.10)$$

D.3.2 Comparison of QED Vacuum Polarization Exponent with Universe Vacuum Polarization Exponent

Eq. D.10 shows the numeric values of the g powers are approximately equal up to a factor of -2. The QED exponent describes high energy vacuum polarization behavior. The universe power g describes the small time universe expansion (near the Big Bang). The relation between the values of g and g_{QED} clearly suggests an analogy.

Further the low energy (infrared) behavior of the QED vacuum polarization which is mass dependent is analogous to the large time (recent time) behavior of a(t) which is governed by the h term in the exponent of a(t).

We now show the universe vacuum polarization is due to a new vector interaction between universes, and show that it is related to the QED vacuum polarization by eq. D.10.[103]

D.3.3 A New Vector Interaction for Universe Particles

We assume universes can be treated as particles in 4-dimensional space-time.[104] Since experiments appear to have shown that our universe does not rotate (does not

---

[101] Eq. 12 in S. Blaha, Phys Rev **D9**, 2246 (1973).

[102] The author calculated α = 1/137… exactly in Blaha (2019a) and (2019b).

[103] The following subsections appeared in Blaha (2019c).

[104] Universes are composite entities but we can treat them as quantum particles in the same manner as physicists treated protons and neutrons etc. as quantum particles before quark theory was accepted. See Blaha (2018e) for a detailed discussion of universe particles.

have spin)[105] we will assume the universe is a spin 0 boson. We assume that universes have a vector field interaction similar to QED.

Given this QED-like framework, then universe-antiuniverse pair production and vacuum polarization becomes possible. We assume the QED-like boson lagrangian

$$\mathcal{L} = \tfrac{1}{2}\,(\partial_\mu\varphi^\dagger\partial^\mu\varphi - m^2\varphi^\dagger\varphi) - ie_0{:}\,\varphi^\dagger(\overrightarrow{\partial_\mu} - \overleftarrow{\partial_\mu})\,\varphi{:}\,A^\mu + e_0^2{:}A^2{:}\;{:}\varphi^\dagger\varphi{:} + \delta m^2{:}\varphi^\dagger\varphi{:} \tag{D.11}$$

where $\varphi(x)$ is a "charged" quantum universe scalar particle field[106] and A^μ is a QED-like field. We now proceed to calculate the second order vacuum polarization of a universe particle. We will assume the term in $\mathcal{L}$ linear in A^μ is the relevant term since the quadratic term always is negligible compared to the linear term in each order α^n of perturbation theory by a factor of α. The neglected terms will be assumed to not affect the calculated eigenvalue function.

D.3.4 Second Order Vacuum Polarization of a Scalar Universe Particle

The one loop vacuum polarization Feynman diagram is

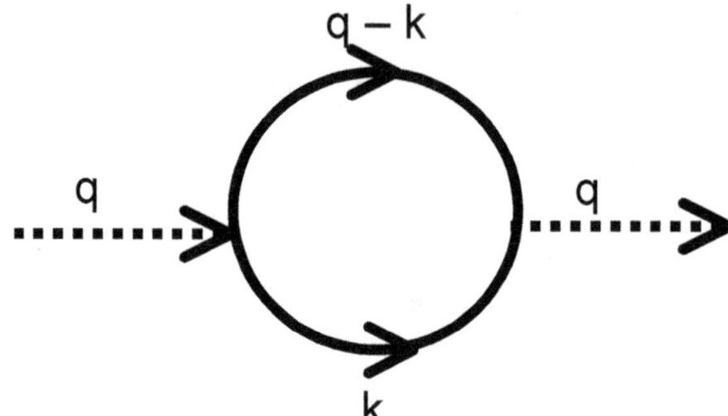

Figure D.1 One loop vacuum polarization boson Feynman diagram.

Its evaluation is

$$I_{\mu\nu} = (-ie_0)^2 \int \frac{d^4k}{(2\pi)^4}\, \frac{i}{(k^2 - m^2 + i\varepsilon)}\, \frac{i}{(k^2 - m^2 + i\varepsilon)}\,(q - 2k)_\mu(q - 2k)_\nu \tag{D.12}$$

$$= \frac{\alpha}{2\pi} \int_0^\infty dz_1 \int_0^\infty dz_2\, \frac{g_{\mu\nu}\exp[i(q^2 z_1 z_2/(z_1 + z_2) - (m^2 + i\varepsilon)(z_1 + z_2))]}{(z_1 + z_2)^3} + \text{gauge terms}$$

upon introducing parameters z_1 and z_2 to enable exponentiation and integration over k, where

---

[105] The lack of universe rotation (spin) is indicated by a study of Cosmic Microwave Background (CMB) by D. Saadeh *et al*, Phys. Rev. Lett. **117**, 313302 (2016).

[106] The charge is not electromagnetic charge.

$$\alpha = e_0^2/4\pi \tag{D.13}$$

After some steps we found

$$I_{\mu\nu} = \frac{i\,\alpha}{12\pi} q^2 g_{\mu\nu} \ln(\Lambda^2/m^2) + \ldots \tag{D.14}$$

with finite and other gauge terms not shown.

Thus we find the renormalization constant Z_{3U} for a scalar universe particle is

$$Z_{3U} = 1 - \alpha/12\pi \ln(\Lambda^2/m^2) \tag{D.15}$$

If we let

$$\alpha_U = \alpha/4 \tag{D.16}$$

then we obtain the form similar to the one loop value of Z_3 for spin ½ electron QED:

$$Z_{3U} \cong 1 - \alpha_U/3\pi \ln(\Lambda^2/m^2) \tag{D.17}$$

We now *provisionally assume* that α is the QED fine structure constant. We denote it as $\alpha_{QED.}$ We verify this choice later.

Thus the "fine structure constant" α_U for our vector interaction is

$$\alpha_U \equiv \alpha_{QED}/4 = 0.001824338 \tag{D.18}$$

We now turn to the Johnson-Baker-Willey (JBW) model of massless QED since at ultra-high energy our vector interaction theory with lagrangian eq. D.11 becomes the JBW model for a scalar particle. In the JBW model we calculated α_{QED} and found the corresponding power of the Z_3 divergent factor which we denote g_{QED}.

D.3.5 Finding the Universe g_U

Now we perform the same calculation for universe vacuum polarization and find the g value, which we denote g_U, corresponds to α_U. The value of g_U will be seen to lead to the power g in the universal scale factor almost exactly.

The universe eigenvalue function is[107]

$$F_2(\alpha_U) = F_1(\alpha_U) - [2/3 + \alpha_U/(2\pi) - (1/4)[\alpha_U/(2\pi)]^2] \tag{D.19}$$

For

$$\alpha_U \equiv \alpha_{QED}/4 = 0.001824338 \tag{D.20}$$

we found the eigenfunction value

$$F_2(\alpha_U = 0.001824338) = 5.10824 \times 10^{-12} \cong 0 \tag{D.21}$$

---

[107] We assume the universe eigenvalue function has the same form as the QED eigenvalue function.

Examining $F_2(\alpha_U)$[108] as a function of g_U we found the value of g_U corresponding to α_U is

$$g_U = -0.00014525 \qquad \text{(D.22)}$$

Thus the universe vacuum polarization is

$$\Gamma_U(p) = (p/\Lambda)^{2g_U} \qquad \text{(D.23)}$$

The fourier transform is[109]

$$a(t) = (1/2\pi)\int_0^\infty dp/p \exp(-ipt)\, \Gamma_U(p) \qquad \text{(D.24)}$$

$$= k\,(t/T)^{-2g_U} \qquad \text{(D.25)}$$

where k is a constant and where

$$1/T = \Lambda \qquad \text{(D.26)}$$

with Λ being the "momentum space" cutoff mass. Comparing to D.10 we find

$$\begin{aligned} g &= -2g_U \\ &= 0.0002905 \end{aligned} \qquad \text{(D.27)}$$

From eq. D.25 for the power g of a(t) we see the universal scale factor g is

$$g = 0.000282377 \qquad \text{(D.28)}$$

Thus the value of g calculated from the universe vacuum polarization differs from the actual value of g by less than 3%. Given the approximate nature of our JBW calculation of vacuum polarization the agreement is remarkable.[110]

In addition we found the "fine structure constant" for the vector interaction to be given by eq. D.16 resulting in

$$e_U = (4\pi\alpha_U)^{1/2} = 0.151411 \qquad \text{(D.29)}$$

Thus we have shown the universe vacuum polarization $\Gamma_U(p)$ when transformed to time is the universal scale factor a(t) up to a constant. The evolution

---

[108] $F_2(\alpha_U)$ and $F_2(g_U)$ are alternate notations for the same function.

[109] Those who might object to fourier transforming to time t should remember that inside a Black Hole the "time-like" coordinate is the radius and the time variable t is comparable to a spatial coordinate. The possibility that the universe is a Black Hole is not excluded. This fourier transform appears in Blaha (2019c) in eq. 25.25 with a typographic error—the division by p was omitted.

[110] And may be exact! The value of the Hubble Constant H in recent times varies from about 70 – 75 making the calculation of g also approximate. We chose an average value of 73.24 to obtain the value of g above. If we chose the current value for H to be 75.58 we would have $g = -2g_U$ exactly. Note: studies of binary black hole merger gravity waves have given a Hubble Constant of 75.2 km s^{-1} Mpc^{-1} (and earlier of 78 km s^{-1} Mpc^{-1}), and studies of light bent by distant galaxies give H = 72.5 km s^{-1} Mpc^{-1}. Thus the value H = 75.58 is not unreasonable.

of our universe is set by universe vacuum polarization. Other 4D universes may be expected to be similar.

The above relation we have found between QED-like vacuum polarization and universe vacuum polarization (Dark Energy) appears to confirm our interpretation of universe Dark Energy as mainly a consequence of universe vacuum polarization due to a universe vector interaction.[111]

D.3.6 Dark Energy is Equivalent to Universe Vacuum Polarization

Dark Energy is elusive both on the experimental and theoretical levels. We know it exists through its effects on our universe. Yet interactions with matter have not been found. Thus it is somewhat of a phantom.

The existence of Dark Energy, which, clearly, strongly affects the evolution of the universe, means that the Einstein equation, usually regarded as central to universe evolution, is incomplete for that purpose. It does not specify the total energy density ρ_{tot}.

$$\dot{a}^2 - 8\pi G\rho_{tot}a^2/3 = -k \tag{D.30}$$

However we can obtain a "handle" on the total energy density by inserting our universal scale factor a(t) in the Einstein equation together with the known radiation density, matter density and Cosmological Constant Λ terms:

$$\rho_{tot}(t) \equiv \rho_{crit}\Omega_{tot}(t) = \rho_{crit}[\Omega_\Gamma(t) + \Omega_M(t) + \Omega_\Lambda + \Omega_T]$$

where the unknown part needed to makes the Einstein equation correct is the elusive Dark Energy $\rho_T(t)$

$$\rho_T(t) = \rho_{crit}\Omega_T(t) \tag{D.31}$$

D.3.7 Quasi-Free Universe Particles

Since $F_2 \cong 0$, universe particles are very much like free particles. (The vacuum polarization is effectively zero.

Universe particles are not totally free particles due to gravitation and Standard Model interactions such as electromagnetism. We treated the case of free universe particles in Blaha (2018e).

D.3.8 Doubling Relation Between Coupling Constants

The coupling constants that we have derived show a doubling whose fundamental significance remains to be understood.

---

[111] Rather like the discovery of the Ω^- particle in the 1960s confirmed Gell-Mann's SU(3) theory.

INTERACTION	COUPLING CONSTANT[112]
Universe Interaction e_U	0.1514
QED $e_{QED} = (4\pi\alpha_{QED})^{1/2}$	0.303
Weak SU(2) g_W	0.619
Strong SU(3) g_S	1. 22

Figure D..2 The interaction coupling constants show a regular doubling. A fundamental cause for doubling is not apparent.

[112] M. Tanabashi *et al* (Particle Data Group), Phys. Rev. D**98**, 030001 (2018).

Appendix E. Some Features of UTMOST

This appendix describes the 64 complex octonion dimension space of UTMOST. It also discusses an alternate 32 *quaternion octonion*[113] dimension space for UTMOST. Given the present state of physical data these variants are both acceptable. and features of QUeST.

E.1 UTMOST Space

An octonion contains eight dimensions. A complex octonion contains sixteen dimensions. A *quaternion octonion*[114] contains 32 dimensions. Fig. E.1 depicts the 64 dimension complex octonion space as a 64 × 16 array of dimensions. It uses a "dot" or pebble • to represent a dimension[115]. The dimensions of the space are not assigned physically until they are mapped to internal symmetry group fundamental representation dimensions and space-time dimensions. Rather than create a cumbersome coordinate-based notation we choose to use •'s.

Figure E.1. The 64 complex octonion dimension UTMOST array. This is the 64 × 16 array of •'s. It has 1024 dimensions.

The UTMOST space can also be viewed as a space of 32 quaternion octonion dimensions. It also has a total of 1024 dimensions. Fig. E.2 creates a 32 × 32 array of dimensions for this space. *A 32 × 32 array is important for the derivation of UTMOST from one-dimension BMOST.*

The 32 × 32 form of the UTMOST dimension array is based on a 32 quaternion octonion dimension space. The difference between this form of space and the 64 complex octonion dimension space above is not physically meaningful at present. The difference will be physically meaningful if the masses of the fermion spectrum and the

---

[113] A quaternion octonion is a 32 dimension set of coordinates. It is the composition of a quaternion and an octonion that could be represented by an expression such as q(o), which represents an octonion argument o for a quaternion functional q that expands each dimension in the octonion fourfold It is a generalization of a complex octonion..

[114] This notation, originated here, is for the composition of a quaternion and an octonion.

[115] The use of pebbles is a useful simplification of coordinates that enables assignments of pebbles (coordinates) to group representations to be visual.

full pattern of symmetry breaking are determined. Then one can differentiate between the symmetry group spectrum and mass spectrums of the respective possibilities.

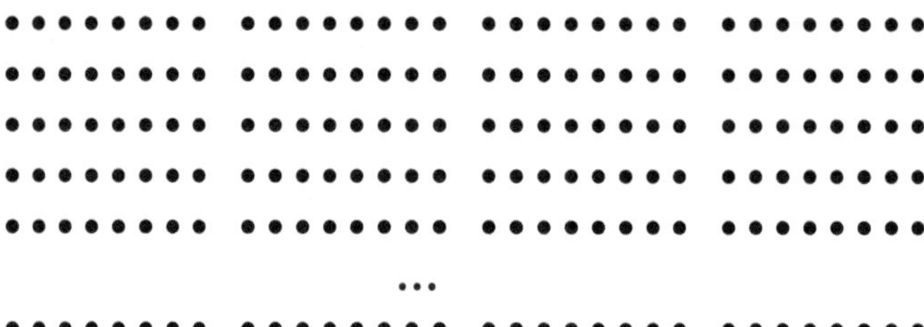

Figure E.2. The UTMOST array with 32 × 32 dimensions for a 32 quaternion octonion dimension space.

The repetitive pattern of groups seen in QUeST leads us to assume that UTMOST has a similar repetitive pattern. We will use a four layer format for the 32 × 32 array of dimensions. Each layer consists of 8 rows of Fig. E.2. Each layer can be put in a form analogous to Fig. C.4 (and to Fig. C.9). See Fig. E.3.

We map between dimensions and fundamental group representations. We use the maps in Table E.1 to set up the group ↔ dimension map, bearing in mind the group representations of the Standard Model:

U(4)	↔	8 real dimensions
U(2)	↔	4 real dimensions
SU(3)	↔	6 real dimensions
U(1)⊗SU(2)	↔	4 real dimensions

Table E.1. Map between fundamental representations and their dimensions.

Fig. E.3 shows the content of one UTMOST layer. The four layers of UTMOST are four copies of Fig. E.3.[116] The separation of the set of dimensions is accomplished by following the procedure given earlier.[117]

Fig. E.5 shows the four layers (each in two rows) of the 32 × 32 dimension UTMOST array, which is composed of 4 × 4 blocks. The 4 × 4 blocks are within the four block 8 × 8 sections for each pair: Normal+Dark1, Dark2+Dark3, Dark4+Dark5 and Dark6+Dark7. In total they form the 32 × 32 = 1024 UTMOST dimension array.

[116] The Layer groups are U(4) groups. They mix the generations of each of the top four layers, generation by generation, separately from the Layer groups mixing the lower four layers. This feature enables QUeST universes to be generated from either the top four layers or the lower four layers.

[117] The separation of the dimensions into the subgroup factors' representation can be implemented as group transformations and definitions using standard group theoretic methods. A more formal method for extracting the subgroup content of representations uses a symmetric group analysis of U(n) representation characters. See S. Blaha, J. Math. Phys. **10**, 2156 (1969) for a detailed discussion of this approach.

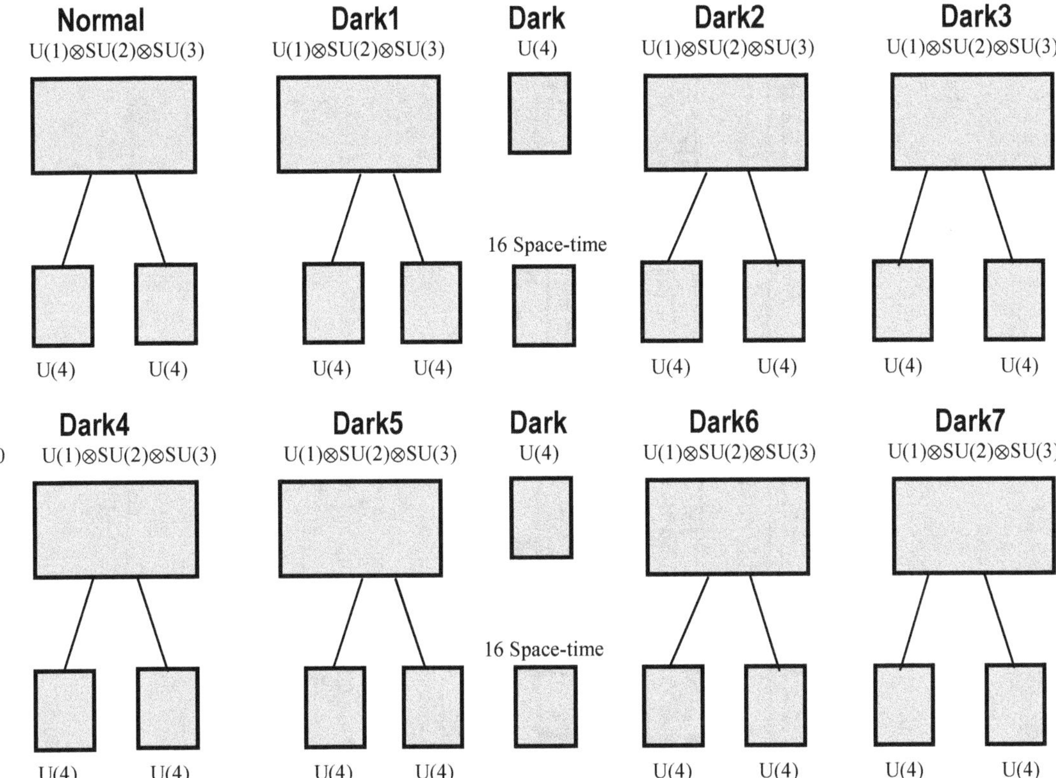

Figure E.3. The internal symmetry groups of *one layer* (consisting of 8 rows in Fig. E.2) of the four layers of 32 $\times$ 32 dimension UTMOST. The other three layers are copies of the this layer. Note the Dark U(4) groups. One U(4) "rotates" among Normal, Dark1, Dark2, and Dark3. The other U(4) "rotates" among Dark4, Dark5, Dark6, and Dark7. Thc U(4) group may be broken to U(1)⊗U(1)⊗U(1)⊗U(1) to obtain 4 $\times$ 4 blocks.

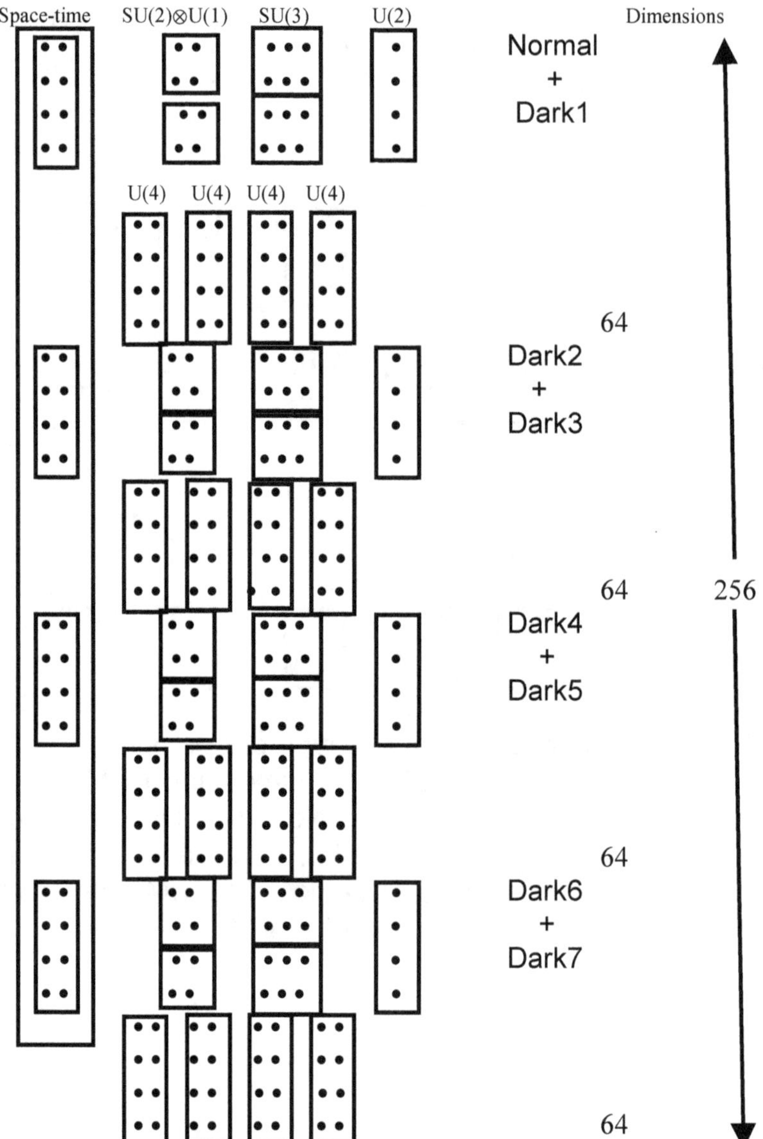

Figure E.4 The *first* of the four layers of UTMOST dimensions with boxes around sets of dimensions for fundamental group representations. The U(4) Dark groups have been broken into U(2)⊗U(2) factors. The U(4) group may be further broken to U(1)⊗U(1)⊗U(1)⊗U(1) to obtain 4 × 4 blocks.

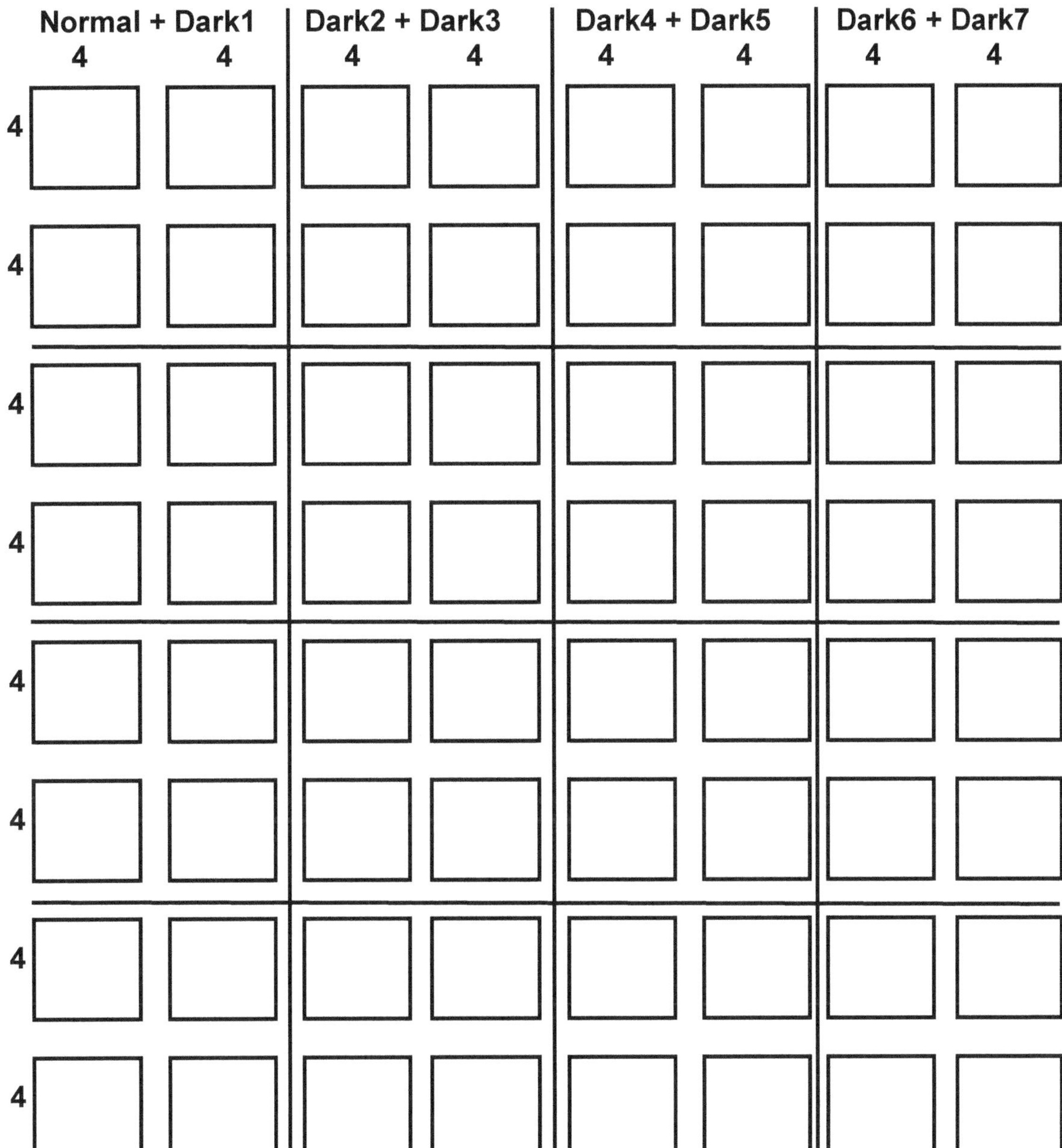

Figure E.5. Four layers (each in two rows) in the 32 × 32 dimension UTMOST array composed of 4 × 4 blocks, which are within the four block 8 × 8 sections for each pair: Normal+Dark1, Dark2+Dark3, Dark4+Dark5 and Dark6+Dark7. In total they form the 32 × 32 = 1024 UTMOST dimension array.

E.2 UTMOST Fermions

Given the form of the internal symmetries in UTMOST we can determine the fermions in the fundamental group representations as shown in Fig. E.6.

UTMOST Fermion Array

Normal	Dark1	Dark2	Dark3	Dark4	Dark5	Dark6	Dark7
••••••••	••••••••	••••••••	••••••••	••••••••	••••••••	••••••••	••••••••
••••••••	••••••••	••••••••	••••••••	••••••••	••••••••	••••••••	••••••••
••••••••	••••••••	••••••••	••••••••	••••••••	••••••••	••••••••	••••••••
••••••••	••••••••	••••••••	••••••••	••••••••	••••••••	••••••••	••••••••
••••••••	••••••••	••••••••	••••••••	••••••••	••••••••	••••••••	••••••••
••••••••	••••••••	••••••••	••••••••	••••••••	••••••••	••••••••	••••••••
••••••••	••••••••	••••••••	••••••••	••••••••	••••••••	••••••••	••••••••
••••••••	••••••••	••••••••	••••••••	••••••••	••••••••	••••••••	••••••••
••••••••	••••••••	••••••••	••••••••	••••••••	••••••••	••••••••	••••••••
••••••••	••••••••	••••••••	••••••••	••••••••	••••••••	••••••••	••••••••
••••••••	••••••••	••••••••	••••••••	••••••••	••••••••	••••••••	••••••••
••••••••	••••••••	••••••••	••••••••	••••••••	••••••••	••••••••	••••••••
••••••••	••••••••	••••••••	••••••••	••••••••	••••••••	••••••••	••••••••
••••••••	••••••••	••••••••	••••••••	••••••••	••••••••	••••••••	••••••••
••••••••	••••••••	••••••••	••••••••	••••••••	••••••••	••••••••	••••••••
••••••••	••••••••	••••••••	••••••••	••••••••	••••••••	••••••••	••••••••

Figure E.6. Spectrum of UTMOST fermions in a 16×64 format. Each fermion is represented by a •..Each set of eight •.'s represents a charged lepton, a neutral lepton, three up-type quarks, and three down-type quarks. There are eight sets of four species in four generations which are in turn in 4 layers. There are 1024 fundamental fermions taking account of quark triplets.

In chapter 3 we outlined possible patterns of subspaces of QUeST and UTMOST. One choice of pattern is based on 4 × 4 blocks of dimensions, assembled into 8 × 8 blocks of dimensions containing four 4 × 4 blocks, assembled in four layers. Fig. E.7 shows the possible implications of this arrangement for fermions. The 4 × 4 fermion blocks contain either four generations of charged leptons and up-quarks, or four generations of neutral leptons and down-quarks.

The grouping of a lepton and three quarks in both cases creates a similarity to time and spatial coordinates respectively suggesting a broken Lorentz group-like structure or a possible SU(4) broken symmetry.

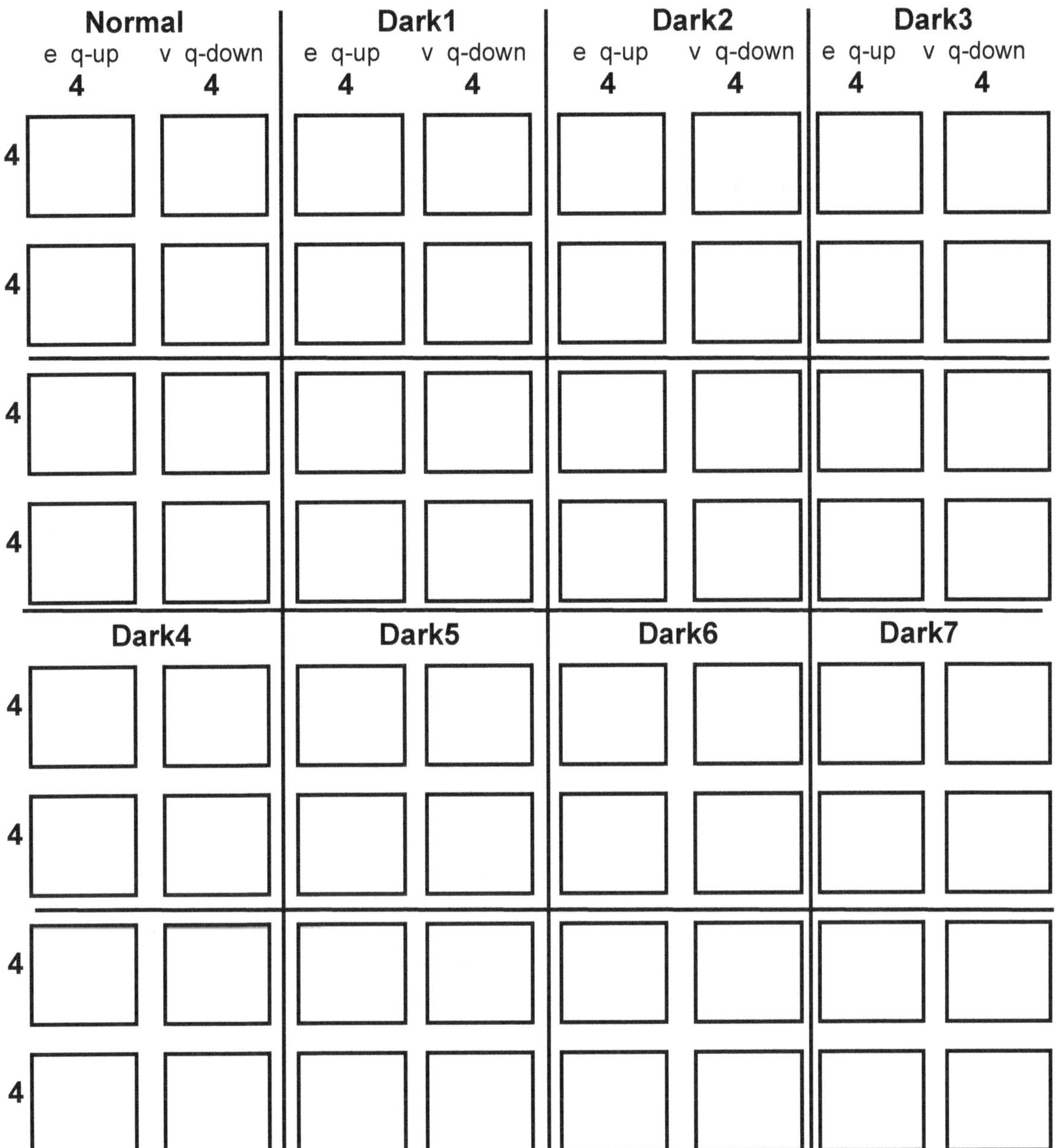

Figure E.7. Block form of the 32 × 32 UTMOST fermion array with each row corresponding to *half of an UTMOST layer*. Thus 8 × ½ = 4 layers results. Each block contains four generations of fermions. The result is sixty-four 4 × 4 blocks. The label e q-up indicates a charged lepton – up-type quark pair, v q-down indicates a neutral lepton – down-type quark pair, and so on. The form displayed here may explain why generations come in fours.

E.3 Partition of UTMOST into QUeST Subspaces

UTMOST[118] can be partitioned into several levels of subspaces. We modify the UTMOST figures seen earlier to show a partition into two subspaces. These spaces will be MOST spaces.[119] Then we can partition a MOST subspace into two QUeST subspaces and so on.

The partitioned UTMOST dimension arrays are:

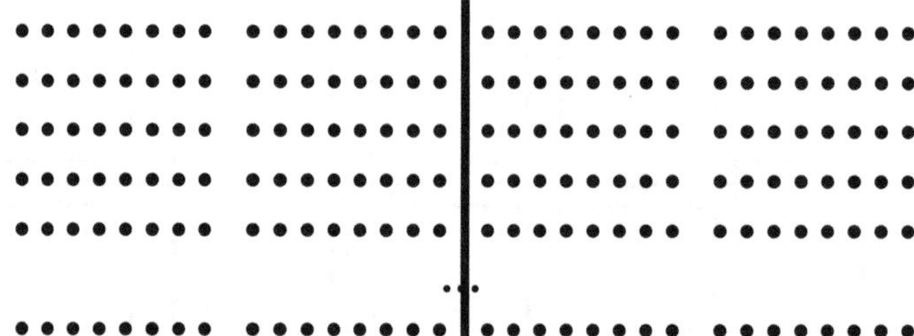

Figure E.8. The partition of the 32 × 32 dimension UTMOST array into MOST subspaces The size of each subspace is 32×16 = 512 dimensions.

The partition of UTMOST fermions into MOST fermions appears in Fig. E.9.

[118] We will consider MOST subspaces below.

[119] MOST is described in Blaha (2020i) and earlier books by the author.

UTMOST Fermion Array Partitioned into MOST Fermion Arrays

Normal	Dark1	Dark2	Dark3	Dark4	Dark5	Dark6	Dark7
••••••••	••••••••	••••••••	••••••••	••••••••	••••••••	••••••••	••••••••
••••••••	••••••••	••••••••	••••••••	••••••••	••••••••	••••••••	••••••••
••••••••	••••••••	••••••••	••••••••	••••••••	••••••••	••••••••	••••••••
••••••••	••••••••	••••••••	••••••••	••••••••	••••••••	••••••••	••••••••
••••••••	••••••••	••••••••	••••••••	••••••••	••••••••	••••••••	••••••••
••••••••	••••••••	••••••••	••••••••	••••••••	••••••••	••••••••	••••••••
••••••••	••••••••	••••••••	••••••••	••••••••	••••••••	••••••••	••••••••
••••••••	••••••••	••••••••	••••••••	••••••••	••••••••	••••••••	••••••••
••••••••	••••••••	••••••••	••••••••	••••••••	••••••••	••••••••	••••••••
••••••••	••••••••	••••••••	••••••••	••••••••	••••••••	••••••••	••••••••
••••••••	••••••••	••••••••	••••••••	••••••••	••••••••	••••••••	••••••••
••••••••	••••••••	••••••••	••••••••	••••••••	••••••••	••••••••	••••••••
••••••••	••••••••	••••••••	••••••••	••••••••	••••••••	••••••••	••••••••
••••••••	••••••••	••••••••	••••••••	••••••••	••••••••	••••••••	••••••••
••••••••	••••••••	••••••••	••••••••	••••••••	••••••••	••••••••	••••••••
••••••••	••••••••	••••••••	••••••••	••••••••	••••••••	••••••••	••••••••

Figure E.9. Partition of spectrum of UTMOST fermions in a 16×64 format. Each fermion is represented by a •. Including each quark. Each set of eight •.'s represents a charged lepton, a neutral lepton, three up-type quarks, and three down-type quarks. There are eight sets of four species in four generations which are in turn in 4 layers. There are 512 fundamental fermions in each subspace taking account of quark triplets. Note: Quark singlets won't do; triplets are required.

UTMOST Fermion Array Partitioned into MOST Fermion Arrays with 4 × 4 Blocks

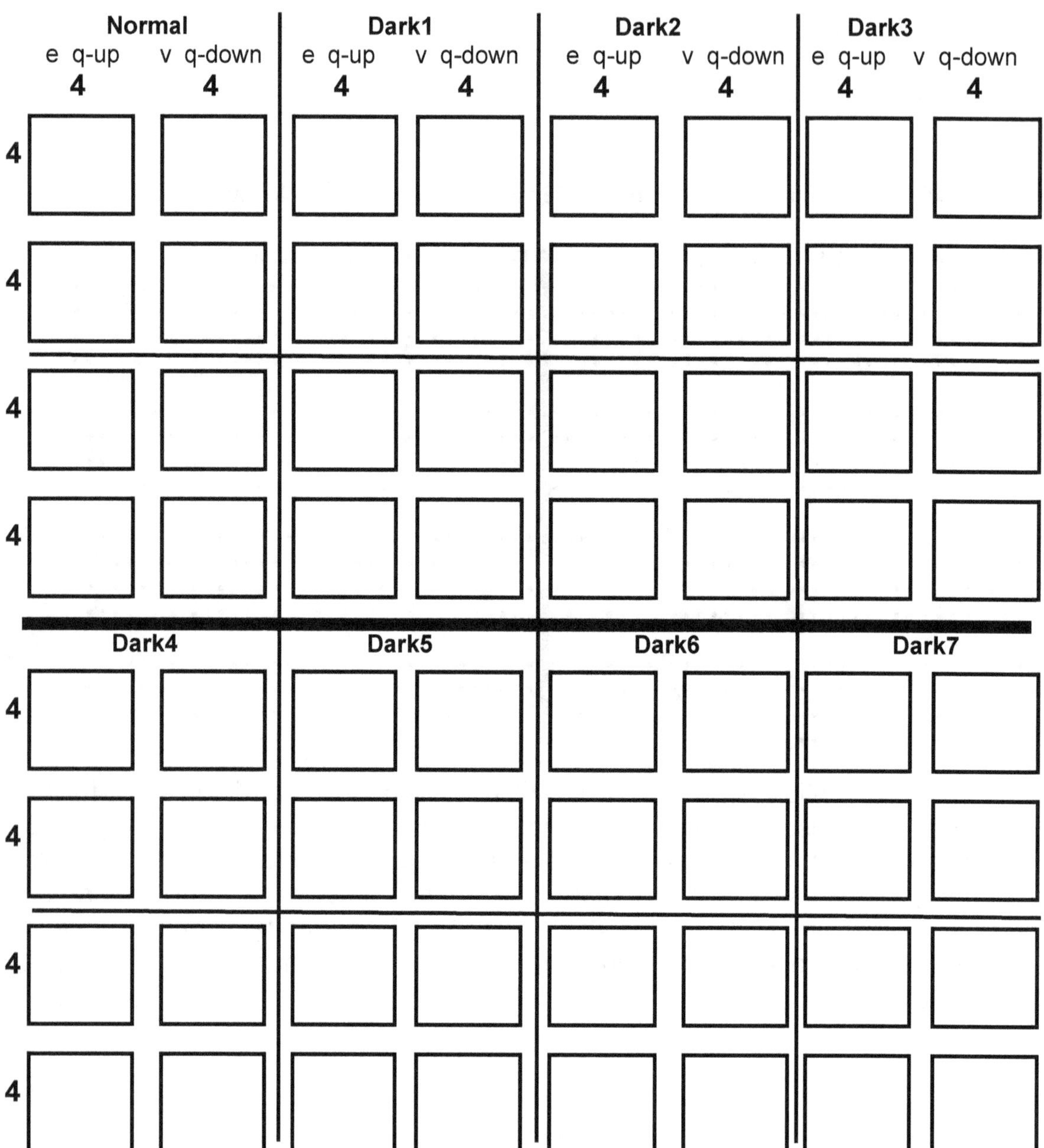

Figure E.10. Partition of block form of the 32 × 32 UTMOST fermion array with each row corresponding to *half of an UTMOST layer*. Thus 8 × ½ = 4 layers results. Each block contains four generations of fermions. The result is sixty-four 4 × 4 blocks. The label e q-up indicates a charged lepton – up-type quark pair, v q-down indicates a neutral lepton – down-type quark pair, and so on.

Appendix F. "D" Dimension Quantum Field Theory from Chapter 10 of Blaha (2017c)

This appendix describes D dimension Quantum Field Theory. It appeared in chapter 10 of Blaha (2017c). The case of the BQUeST seed fermion requires dimension D = 8. The seed urfermion for BMOST requires D = 10. A 16 dimension quantum field theory version appears in Blaha (2014a).

10.2 The Particle Interpretation of Extended Wheeler-DeWitt Equation Solutions

In earlier chapters we described features of the Wheeler-DeWitt equation that suggested that universes could be viewed as particles or anti-particles, or tachyons. The solutions of this equation are scalar wave functions on a manifold that are analogous to the solutions of the Klein-Gordon equation. The issues of negative probabilities, possible tachyonic solutions, and negative frequency solutions suggest a need for an appropriate particle interpretation of universes that can possibly resolve these problems.

Some physicists have taken the Wheeler-DeWitt equation as the starting point for a theory of a universe as a particle. The Wheeler-DeWitt equation describes the interior of a universe in a quantum framework.

We will take a different approach using the Megaverse as the environment of universe particles that internally have Quantum Gravity, and externally have Megaverse Quantum Gravity.

We view a universe as an extended particle and begin by ignoring the detailed inner structure of universes. This approach is similar to the historical treatment of hadrons such as the proton as particles and developing a theory of them as fundamental particles using form factors, structure functions and so on to approximate their inner structure. Afterwards, as detailed data became available, the detailed investigation of the internal structure of hadrons using quark-parton models followed. We will pursue a similar theoretical development beginning with a theory of universes as extended particles in the D-dimension Megaverse. The internal structure of the particle universes will eventually be specified by the Wheeler-DeWitt equation expressed in Megaverse coordinates.

The two simplest choices for the nature of universes are "spin 0" *bosonic universes* and *fermionic universes* with odd half integer spin, s_M.[120] We will first consider the possibility of fermionic universes, and then briefly consider "spin 0" *bosonic* universes.

The first issue of fermionic universes (reminiscent of the discussions of spin in the 1920's) is the interpretation of spin states. We suggest that the upper $2^{D/2\ -\ 1}$

[120] Since the Megaverse is D-dimensional, the spin of fermionic universe particles was shown to be s_M.

components (with $2^{D/2-2}$ "spin up" and $2^{D/2-2}$ "spin down" states) of a fermionic universe wave function represent a left-handed universe with an excess number of baryons. The lower ($2^{D/2-1}$) components lead to right-handed anti-universes where there is an excess of anti-baryons. These associations are analogous to the interpretations of the Dirac electron wave function.[121]

The universe particle "spin up" and "spin down" states are distinguished by their interactions with gauge fields in a manner analogous to quantum electrodynamics.

10.3 "Free Field" Dynamics of Fermionic Universe Particles

We now consider universes as extended particles with an odd half integer spin – *fermionic universe particles* - in the D-dimensional Megaverse. In the Megaverse there are D 'Dirac' matrices with $2^{D/2}$ rows and $2^{D/2}$ columns that are the equivalent of the four Dirac matrices in four dimensions. We will denote these D matrices as $\gamma_M{}^i$ for i = 1, 2, … , D. They satisfy the anti-commutation relations:

$$\{\gamma_M{}^i, \gamma_M{}^j\} = 2\,\delta^{ij} \tag{10.1}$$

and thus form a Clifford algebra. We will choose y^D to be the time coordinate and thus make it pure imaginary with a Reality group transformation. (The D-dimensional Megaverse space is a complex Euclidean space.) Therefore γ^D will be hermitean ($(\gamma^D)^2 = 1$), and the γ^i matrices for i = 1, … , (D – 1) will be anti-hermitean with $(\gamma^i)^2 = -1$. The number of linearly independent matrices in D dimensions is 2^D.

The Megaverse metric is (by use of the Reality group) chosen to be

$$g^{ij} = -\delta^{ij}, \qquad g^{D,D} = 1 \tag{10.2}$$

for i, j = 1, 2, … , (D – 1); and zero otherwise.

Except for the additional dimensions, fermion dynamics is quite similar to the 4-dimensional case. The free universe particle Dirac equation is

$$(i\gamma^i\partial_i + m)\psi(y) = 0 \tag{10.3}$$

summed over i = 1, 2, … , D where the mass is assumed to be constant, and set by eq. 10.119 below. The derivative operator, is based on the use of quantum coordinates[122] (eq. 7.9)

$$Y^i(y) = y^i + i\,Y_u{}^i(y)/M_u{}^{D/2} \tag{10.4}$$

For i = 1, …, D and is defined to be

---

[121] It is known that phenomena in our universe tend to be left-handed. If this feature of our universe's phenomena is also a property of the universe itself, then, since handedness is an attribute of spin, the treatment of a universe as having spin is not unreasonable.

[122] Giving Two Tier renormalization.

$$\partial_i = \partial/\partial Y^i(y) = \partial/\partial(y_i - Y_{ui}(y)/M_u^{D/2}) \tag{10.5}$$

where *we assume $M_u = M_c$ with M_c being a very large mass scale of perhaps the order of the Planck mass.*

Y_u^i is a D-dimensional Megaverse gauge field equivalent of the universe $Y^\mu(x)$ used in Two Tier renormalization (discussed in I):

$$Y^\mu(z) = z^\mu + i\, Y^\mu(z)/M_c^2$$

where $Y^\mu(z)$ is a free QED-like field. The $Y^i(y)$ quantum coordinates will be used in the Megaverse to eliminate potential divergences, in a manner similar to the case of our universe when universe particle interactions are introduced later.

10.3.1 Four Types of Fermionic Universe Particles

Assuming universe energies are real-valued,[123] there are four possible types of fermionic universe particles in the Megaverse that are analogous to the four species of fermion described in I (and Blaha (2010b)) for The Extended Standard Model. Two of these types are tachyonic. It is important to note that DeWitt points out that the Wheeler-DeWitt equation has tachyonic solutions since the mass-like term dependent on ${}^{(3)}R$ can be positive or negative.[124] A negative mass is an indication of tachyonic behavior wherein the wave propagation of the state functional is not necessarily in time-like directions and is thus tachyonic.

Eq. 10.3 is a Dirac-type D-dimensional Dirac equation. There are three other general types of universe particle equations. (By assumption fermionic universes come in four species like fermions.) The derivation of the four types of universe particles is similar to the derivation of fermion types in the Extended Standard Model in 4-dimensional complex space-time given in Blaha (2010b). We will now consider the D-dimensional equivalent for universe particles in the Megaverse.

The general form of a pure D-dimensional complex Lorentz group[125] boost can be expressed in terms of a complex relative (D – 1)-velocity $\mathbf{v_c}$ between inertial reference frames. A D-dimensional coordinate boost has the form

$$\Lambda_C(\mathbf{v_c}) \equiv \Lambda_C(\omega, \mathbf{v_c}) = \exp[i\omega\hat{\mathbf{w}}\cdot\mathbf{K}] \tag{10.6}$$

where

$$\omega = (\omega_r^2 - \omega_i^2 + 2i\omega_r\omega_i\, \hat{\mathbf{u}}_r\cdot\hat{\mathbf{u}}_i)^{1/2} \tag{10.7}$$

and

[123] The energy of universe particles need not be real-valued since universes can 'decay' – unlike elementary particles which are not subject to decay, by definition, since they are assumed to be *fundamental*. We choose to consider the case of universes with real-valued energies. The case of universes with complex-valued energies is a simple extension of the real-value cases considered here.

[124] DeWitt, B. S., Phys. Rev. **160**, 1113 (1967) p. 1124.

[125] The D-dimensional complex Lorentz group has similar features to the 4-dimensional complex Lorentz group. We shall only discuss it to the extent needed for our universe particle type's derivation. See Weinberg (1995) for the 4-dimensional Lorentz group – the D-dimensional Lorentz group generalizes directly from the features of the 4-dimensional Lorentz group.

$$\hat{\mathbf{w}} = (\omega_r\hat{\mathbf{u}}_r + i\omega_i\hat{\mathbf{u}}_i)/\omega \tag{10.8}$$

with all vectors being (D – 1)-dimensional spatial vectors. We define the real and imaginary unit vectors $\hat{\mathbf{u}}_r\cdot\hat{\mathbf{u}}_r = 1 = \hat{\mathbf{u}}_i\cdot\hat{\mathbf{u}}_i$ with the result

$$\hat{\mathbf{w}}\cdot\hat{\mathbf{w}} = 1 \tag{10.9}$$

The complex relative velocity is

$$\mathbf{v_c} = \hat{\mathbf{w}}\tanh(\omega) \tag{10.10}$$

The free dynamical equations of the four universe particle species will be generated by D-dimensional Lorentz boosts of the free Dirac equation of a universe particle at rest with the *requirement that the time variable* ($t = y^D$) *and energy are real in the resulting field equations.*[126] The procedure can most easily be performed in D-dimensional momentum space with the Megaverse coordinate space version of the generated equation determined from the momentum space version.

10.3.1.1 Dirac-like Equation – Type I universe Particle

A positive energy plane wave solution of the Dirac equation eq. 10.3 for a universe particle at rest is

$$\psi(y) = \exp[-imt]w(0) \tag{10.11}$$

where we set $\partial_t = \partial/\partial y_D$ while temporarily ignoring the $Y_u^{\,i}(y)/M_u^{\,D/2}$ term. w(0) is a $2^{D/2}$ component spinor column vector. The solution ψ(y) satisfies the momentum space Dirac equation for a particle at rest:

$$(m\gamma^D - m)\psi(y) = 0 \tag{10.12}$$

The $2^{D/2} \times 2^{D/2}$ spinor matrix form of a D-dimensional Lorentz boost with a relative real velocity **v** of the Dirac matrices is[127]

$$S^{-1}(\Lambda(\mathbf{v}))\gamma^\nu S(\Lambda(\mathbf{v})) = \Lambda^\nu{}_\mu(\mathbf{v})\gamma^\mu \tag{10.13}$$

where $\Lambda^\nu{}_\mu(\mathbf{v})$ is a D-dimensional Lorentz boost. S(Λ(**v**)) has the form

$$\begin{aligned} S(\Lambda(\mathbf{v})) &= \exp(-\omega\gamma^D\boldsymbol{\gamma}\cdot\mathbf{v}/(2|\mathbf{v}|)) \\ &= \cosh(\omega/2)I + \sinh(\omega/2)\gamma^D\boldsymbol{\gamma}\cdot\mathbf{p}/|\mathbf{p}| \end{aligned} \tag{10.14}$$

---

[126] The D-dimensional “energy” must be real since it relates to the area of the universe – a real number.

[127] **The indices ν and μ from this point in this chapter have values: 1, 2, … , D.**

with *real* $\omega = \text{arctanh}(|\mathbf{v}|)$ and *real* $\mathbf{v}$. $|\mathbf{p}|$ is the magnitude of the spatial (D – 1)-vector. Also

$$S^{-1}(\Lambda(\mathbf{v})) = \gamma^D S^\dagger(\Lambda(\mathbf{v}))\gamma^D = \exp(\omega\gamma^D\boldsymbol{\gamma}\cdot\mathbf{v}/(2|\mathbf{v}|))$$

$$= \cosh(\omega/2)I - \sinh(\omega/2)\gamma^D\boldsymbol{\gamma}\cdot\mathbf{p}/|\mathbf{p}| \tag{10.15}$$

If we now apply $S(\Lambda(\mathbf{v}))$ to the momentum space Dirac equation of a particle at rest (eq. 10.12) we find

$$0 = S(\Lambda(\mathbf{v}))(m\gamma^D - m)\,\psi(y)$$
$$= [mS(\Lambda(\mathbf{v}))\gamma^D S^{-1}(\Lambda(\mathbf{v})) - m]S(\Lambda(\mathbf{v}))w(0)$$

A straightforward evaluation shows

$$mS(\Lambda(v))\gamma^D S^{-1}(\Lambda(v)) = g_{D\mu\nu}p^\mu\gamma^\nu = \not{p} \tag{10.16}$$

where p is a momentum D-vector. In addition we define the D-dimension spinor ($2^{D/2}$ components)

$$S(\Lambda(v))w(0) = w(p) \tag{10.17}$$

which can be viewed as a "positive energy D Dirac spinor". The Dirac equation in momentum space has the familiar form:

$$(\not{p} - m)\exp[-ip\cdot y]w(p) = 0 \tag{10.18}$$

Eq. 10.18 implies the free, coordinate space Dirac equation:

$$(i\gamma^\mu\partial/\partial y^\mu - m)\psi(y) = 0 \tag{10.19}$$

We identify this equation as the dynamical equation of a type 1 universe particle. It corresponds to the free charged lepton elementary particle species Dirac equation in particle physics.

10.3.1.2 Complex Boosts

The form of the D-dimensional spinor boost transformation corresponding to the coordinate transformation eq. 10.6 is:

$$S_C(\omega, \mathbf{v}_c) \equiv S_C = \exp(-\omega\gamma^D\boldsymbol{\gamma}\cdot\hat{\mathbf{w}}/2)$$
$$= \cosh(\omega/2)I + \sinh(\omega/2)\gamma^D\boldsymbol{\gamma}\cdot\hat{\mathbf{w}} \tag{10.20}$$

with *complex* $\mathbf{v}_c$ and $\hat{\mathbf{w}}$ defined by eqs. 10.10 and 10.8 respectively. The inverse transformation is

$$S_C^{-1}(\omega, \mathbf{v_c}) = \exp(\omega\gamma^D\boldsymbol{\gamma}\cdot\hat{\mathbf{w}}/2)$$
$$= \cosh(\omega/2)I - \sinh(\omega/2)\gamma^D\boldsymbol{\gamma}\cdot\hat{\mathbf{w}} \qquad (10.21)$$

Note that S_C is not unitary just as in the 4-dimensional case.

We now apply a spinor boost to the Dirac equation for a particle at rest in this more general case of complex ω and $\hat{\mathbf{w}}$.

$$0 = S_C(\omega, \mathbf{v_c}))(m\gamma^D - m) \exp[-imt]w(0)$$
$$= [mS_C\gamma^D S_C^{-1} - m] \exp[-imt]S_C w(0) \qquad (10.22)$$

where $S_C = S_C(\omega, \mathbf{v_c})$. After some algebra we find

$$mS_C\gamma^D S_C^{-1} = m[\cosh(\omega)\gamma^D - \sinh(\omega)\boldsymbol{\gamma}\cdot\hat{\mathbf{w}}] \qquad (10.23)$$

We will use these *complex* boosts to generate the other species' Dirac-like equations.

10.3.1.3 Tachyon Universe particle Dirac Equation

The development of the complex spinor boost transformation (subsection 10.3.1.2 above) leads to two possible forms of the tachyon Dirac-like equation. One form will lead to a lagrangian dynamics for left-handed universe particles. The other form leads to a lagrangian dynamics for right-handed universe particles.

10.3.1.4 Type IIa Case: Left-Handed Tachyonic Universe Particles

If the real and imaginary relative vectors parts of $\hat{\mathbf{w}}$, namely $\hat{\mathbf{u}}_r$ and $\hat{\mathbf{u}}_i$, are parallel, then $\hat{\mathbf{u}}_r\cdot\hat{\mathbf{u}}_i = 1$ and

$$\omega = \omega_r + i\omega_i \qquad (10.24)$$

Eqs. 10.23 and 10.24 then imply

$$mS_C\gamma^D S_C^{-1} = m[\cosh(\omega_r)\cos(\omega_i) + i\sinh(\omega_r)\sin(\omega_i)]\gamma^D -$$
$$- m[\sinh(\omega_r)\cos(\omega_i) + i\cosh(\omega_r)\sin(\omega_i)]\boldsymbol{\gamma}\cdot\hat{\mathbf{u}}_r \qquad (10.25)$$

or

$$mS_C\gamma^D S_C^{-1} = \cos(\omega_i)\gamma\cdot p_r + i\sin(\omega_i)\gamma\cdot p_i \qquad (10.26)$$

where

$$p_r^0 = m\cosh(\omega_r) \qquad p_i^0 = m\sinh(\omega_r) \qquad (10.27)$$

and

$$\mathbf{p_r} = m\hat{\mathbf{u}}_r\sinh(\omega_r) \qquad \mathbf{p_i} = m\hat{\mathbf{u}}_r\cosh(\omega_r) \qquad (10.28)$$

If $\omega_i = 0$, then we recover the momentum space Dirac-like equation. If $\omega_i = \pi/2$, then we obtain the left-handed momentum space tachyon equation:

$$mS_C\gamma^D S_C^{-1} = i\gamma\cdot p_i \qquad (10.29)$$

and the tachyon energy and momentum expressions

$$\mathbf{p} = m\mathbf{v}\gamma_s \qquad E = m\gamma_s \qquad (10.30)$$

where $\sinh(\omega) = \gamma_s = (\beta^2 - 1)^{-½}$ with $\beta = v/c > 1$. v is the absolute value of the (D – 1) component spatial velocity. Also

$$S_C w(0) = w_C(p) \qquad (10.31)$$

is a tachyon spinor.

The momentum space tachyonic Dirac-like equation is

$$(i\not{p} - m)\exp[-ip\cdot y]w_T(p) = 0 \qquad (10.32)$$

where $p\cdot y = p^D y^D - \mathbf{p}\cdot\mathbf{y}$ after performing a corresponding boost in the exponential factor. If we apply $i\not{p}$ to eq. 10.32 we find the tachyon mass condition is satisfied

$$-E^2 + \mathbf{p}^2 = m^2 \qquad (10.33)$$

Transforming back to coordinate space we obtain the "left-handed" *tachyonic Dirac-like equation*:

$$(\gamma^\mu\partial/\partial y^\mu - m)\psi_T(y) = 0 \qquad (10.34)$$

10.3.1.5 Type IIb Case: Right-Handed Tachyonic Universe Particles

If the real and imaginary relative vectors parts of $\hat{\mathbf{w}}$, $\hat{\mathbf{u}}_r$ and $\hat{\mathbf{u}}_i$, are anti-parallel $\hat{\mathbf{u}}_r = -\hat{\mathbf{u}}_i$, then $\hat{\mathbf{u}}_r\cdot\hat{\mathbf{u}}_i = -1$ and

$$\omega = \omega_r - i\omega_i \qquad (10.35)$$

then

$$mS_C\gamma^D S_C^{-1} = m[\cosh(\omega_r)\cos(\omega_i) - i\sinh(\omega_r)\sin(\omega_i)]\gamma^D - \\ - m[\sinh(\omega_r)\cos(\omega_i) - i\cosh(\omega_r)\sin(\omega_i)]\boldsymbol{\gamma}\cdot\hat{\mathbf{u}}_r \qquad (10.36)$$

or

$$mS_C\gamma^D S_C^{-1} = \cos(\omega_i)\gamma\cdot p_r - i\sin(\omega_i)\gamma\cdot p_i \qquad (10.37)$$

where

$$p_r^{\mathbf{D}} = m\cosh(\omega_r) \qquad p_i^{\mathbf{D}} = m\sinh(\omega_r) \qquad (10.38)$$

and

$$\mathbf{p}_r = m\hat{\mathbf{u}}_r\sinh(\omega_r) \qquad \mathbf{p}_i = m\hat{\mathbf{u}}_r\cosh(\omega_r) \qquad (10.39)$$

If $\omega_i = \pi/2$, then we obtain the right-handed momentum space tachyon equation.[128]

[128] We note that $\gamma_s = (\beta^2 - 1)^{-½}$, *if expressed in terms of* ω, *has a branch cut extending from* <-∞, +∞> *in the complex* ω *plane. Thus values of* ω *with positive imaginary parts are physically different from values of* ω *with negative imaginary parts.*

$$(-\gamma^{\mu}\partial/\partial y^{\mu} - m)\psi_T(y) = 0 \qquad (10.40)$$

10.3.1.6 Type III Case: "Up-Quark-like" Universe Particles

There are two other cases where we can obtain fermion dynamical equations with a *real* time variable and real energy.[129] In one case we set $\hat{\mathbf{u}}_r \cdot \hat{\mathbf{u}}_i = 0$ and have a real ω.

If the real and imaginary relative vectors parts of $\hat{\mathbf{w}}$, namely $\hat{\mathbf{u}}_r$ and $\hat{\mathbf{u}}_i$, are perpendicular, $\hat{\mathbf{u}}_r \cdot \hat{\mathbf{u}}_i = 0$, then

$$\omega = (\omega_r^2 - \omega_i^2)^{½} \qquad (10.41)$$

Thus ω is either pure real ($\omega_r \geq \omega_i$) or pure imaginary ($\omega_r < \omega_i$).

The momentum space equation generated by the corresponding spinor boost is

$$\{m \cosh(\omega)\gamma^D - m \sinh(\omega)\boldsymbol{\gamma}\cdot(\omega_r\hat{\mathbf{u}}_r + i\omega_i\hat{\mathbf{u}}_i)/\omega - m\} \exp[-imt]w_c(p) = 0 \qquad (10.42)$$

Defining the momentum 4-vector

$$p = (p^{\mathbf{D}}, \mathbf{p}) \qquad (10.43)$$

where

$$p^{\mathbf{D}} = m \cosh(\omega) \qquad \mathbf{p} = \mathbf{p}_r + i\mathbf{p}_i \qquad (10.44)$$

with

$$\mathbf{p}_r = m\omega_r\hat{\mathbf{u}}_r \sinh(\omega)/\omega \qquad \mathbf{p}_i = m\omega_i\hat{\mathbf{u}}_i \sinh(\omega)/\omega \qquad (10.45)$$

$$\mathbf{p}_r\cdot\mathbf{p}_i = 0 \qquad (10.46)$$

then we obtain a positive energy Dirac-like equation

$$[p\cdot\gamma - m]\exp[-imt]w_c(p) = 0$$

or

$$[p^{\mathbf{D}}\gamma^{\mathbf{D}} - (\mathbf{p}_r + i\mathbf{p}_i)\cdot\boldsymbol{\gamma} - m]\exp[-ip\cdot y]w_c(p) = 0 \qquad (10.47)$$

with a complex 3-momentum $\mathbf{p}$ and the 4-momentum mass shell condition:

$$p^2 = (p^{\mathbf{D}})^2 - \mathbf{p}_r\cdot\mathbf{p}_r + \mathbf{p}_i\cdot\mathbf{p}_i = m^2 \qquad (10.48)$$

Note

$$|\mathbf{v}_c| = |\mathbf{p}|/p^{\mathbf{D}} = [(\mathbf{p}_r + i\mathbf{p}_i)\cdot(\mathbf{p}_r + i\mathbf{p}_i)]^{½}/p^{\mathbf{D}} = \tanh(\omega) \qquad (10.49)$$

---

[129] The requirement of a real energy for a universe is not strict. For a fundamental free particle the energy must be real or the particle would be subject to decay – contrary to its assumed fundamental nature. Universes can 'decay' to 'smaller' universes. Therefore the requirement for real energy can be violated. Nevertheless the requirement for real energy is appealing since it leads to four species of universes strengthening the analogy of universes to elementary particles.

and so the Lorentz factor is

$$\gamma = \cosh(\omega) \tag{10.50}$$

Eq. 10.47 is the momentum space equivalent of the wave equation[130]

$$[i\gamma^D\partial/\partial t + i\gamma\cdot(\nabla_r + i\nabla_i) - m]\psi_u(t, \mathbf{y}_r, \mathbf{y}_i) = 0 \tag{10.51}$$

where $\mathbf{y} = \mathbf{y}_r - i\mathbf{y}_i$, and where the grad operators ∇_r and ∇_i are with respect to $\mathbf{y}_r$ and $\mathbf{y}_i$ respectively. Since $\hat{\mathbf{u}}_r\cdot\hat{\mathbf{u}}_i = 0$ we see that there is a subsidiary condition on the wave function

$$\nabla_r\cdot\nabla_i\, \psi_u(t, \mathbf{y}_r, \mathbf{y}_i) = 0 \tag{10.52}$$

We note eq. 10.52 can be put into covariant form as the difference of two vectors squared (which is a real D-dimensional Lorentz group invariant):

$$[\gamma^D\partial/\partial t + i\gamma\cdot(\nabla_r + i\nabla_i)]^2 - [\gamma^D\partial/\partial t + i\gamma\cdot(\nabla_r - i\nabla_i)]^2 = 4\nabla_r\cdot\nabla_i.$$

We identify eq. 10.51 as the dynamical equation of an "up-quark-like" universe particle.

10.3.1.7 Type IVa Case: Left-Handed "Down-Quark-like" Tachyonic Universe Particles

In this case we set $\hat{\mathbf{u}}_r\cdot\hat{\mathbf{u}}_i = 0$. Then by eq. 10.7

$$\omega = (\omega_r^2 - \omega_i^2)^{1/2}$$

Thus ω again starts out either pure real (if $\omega_r \geq \omega_i$) or pure imaginary (if $\omega_r < \omega_i$). In this case we also choose ω real, and then change ω to

$$\omega = (\omega_r^2 - \omega_i^2)^{1/2} \rightarrow \omega' = (\omega_r^2 - \omega_i^2)^{1/2} + i\pi/2 = \omega + i\pi/2$$

by adding $i\pi/2$ to ω since ω is a free parameter. We then proceed as we did in the prior tachyon case.[131]. The resulting Lorentz boost

$$\Lambda_C = \exp[i((\omega_r^2 - \omega_i^2)^{1/2} + i\pi/2)(\omega_r\hat{\mathbf{u}}_r + i\omega_i\hat{\mathbf{u}}_i)\cdot\mathbf{K}/\omega] \tag{10.53}$$

becomes a left-handed "quark-like" boost. The tachyon dynamical equation is[132]

---

[130] The gradient operators ∇_r and ∇_i are 15-dimensional spatial gradient operators.

[131] Here again the choice of ω in eq. 10.53 leads to a "left-handed" universe particle while the choice $\omega' = \omega - i\pi/2$ leads to a right-handed one.

[132] The gradient operators ∇_r and ∇_i are (D – 1)-dimensional spatial gradient operators.

$$[\gamma^D\partial/\partial t + \gamma\cdot(\nabla_r + i\nabla_i) - m]\psi_d(y) = 0 \qquad (10.54)$$

with the constraint equation

$$\nabla_r\cdot\nabla_i\ \psi_d(t, \mathbf{y}_r, \mathbf{y}_i) = 0 \qquad (10.55)$$

We will call the universe particles satisfying eqs. 10.54 and 10.55 left-handed *tachyonic quark-like universe particles.*

10.3.1.8 Type IVb Case: Right-Handed Down-Quark-like Tachyonic Universe Particles

In this case we set $\hat{\mathbf{u}}_r\cdot\hat{\mathbf{u}}_i = 0$. Then by eq. 10.7

$$\omega = (\omega_r^2 - \omega_i^2)^{1/2}$$

Thus ω again starts out either pure real (if $\omega_r \geq \omega_i$) or pure imaginary (if $\omega_r < \omega_i$). In this case we also choose ω real, and then change ω to

$$\omega = (\omega_r^2 - \omega_i^2)^{1/2} \rightarrow \omega' = (\omega_r^2 - \omega_i^2)^{1/2} - i\pi/2 = \omega - i\pi/2$$

since ω is a free parameter and proceed as we did in the prior case. The resulting Lorentz boost

$$\Lambda_C = \exp[i((\omega_r^2 - \omega_i^2)^{1/2} - i\pi/2)(\omega_r\hat{\mathbf{u}}_r + i\omega_i\hat{\mathbf{u}}_i)\cdot\mathbf{K}/\omega] \qquad (10.56)$$

becomes a right-handed quark-like boost. The resulting tachyon dynamical equation is

$$[-\gamma^D\partial/\partial t - \gamma\cdot(\nabla_r + i\nabla_i) - m]\psi_d(y) = 0 \qquad (10.57)$$

with the constraint equation

$$\nabla_r\cdot\nabla_i\ \psi_d(t, \mathbf{y}_r, \mathbf{y}_i) = 0 \qquad (10.58)$$

We will call the universe particles satisfying eqs. 10.57 and 10.58 right-handed *tachyonic quark-like universe particles.*

10.3.2 Lagrangians

In this section we will develop a lagrangian formalism for each of the four types of fermionic universe particles noting that a tachyonic universe particles have two forms: left-handed and right-handed (discussed later in section 10.3.5).

The various types of universe particles described in section 10.3.1 correspond to universes with differing internal characteristics and motion in the Megaverse. The equations are all free field equations. Internal potentials and interactions must be introduced in these equations to complete the universe dynamical equations. A connection to the Wheeler-DeWitt description of their internal quantum structure also remains to be established (section 10.3.6).

In defining the lagrangians for the four fermionic universe types that yield their dynamical equations in a canonical manner, we require the conventional quantum field theory feature that the hamiltonian derived from the lagrangian is hermitean. We will develop a separate lagrangian for each type.

10.3.2.1 Type I Universe Particle Lagrangian

The Universe particle Dirac equation lagrangian is

$$\mathcal{L}_u = \bar{\psi}(i\gamma^\mu \partial/\partial y^\mu - m)\psi(y) \tag{10.59}$$

where

$$\bar{\psi} = \psi^\dagger \gamma^D$$

and $\psi^\dagger$ is the hermitean conjugate of ψ.

10.3.2.2 Type II Tachyon Universe Particle Lagrangian

This lagrangian includes both left-handed and right-handed cases. It can be separated into lagrangian terms for each case using parity projection operators.

$$\mathcal{L}_{uT} = \psi_T{}^S(\gamma^\mu \partial/\partial y^\mu - m)\psi_T(y) \tag{10.60}$$

where

$$\psi_T{}^S = \psi_T{}^\dagger\, i\gamma^D\gamma^5 \tag{10.61}$$

with γ^5 being the D-dimensional equivalent for γ^5 in 4 dimensions. The peculiar form of the tachyon universe lagrangian is necessitated by the hermiticity of the hamiltonian calculated from it.

10.3.2.3 Type III "Up-Quark-like" Universe Particle Lagrangian

The lagrangian density of a free "up-quark-like" universe particle is

$$\mathcal{L}_u = \bar{\psi}_u(i\gamma^\mu D_\mu - m)\psi_u(y) \tag{10.62}$$

where $\bar{\psi}_u = \psi_u{}^\dagger\gamma^D$ and

$$\psi_u{}^\dagger = [\psi_u(\mathbf{y}_r, \mathbf{y}_i)]^\dagger \big|_{\mathbf{y}_i = -\mathbf{y}_i} \tag{10.63}$$

$$D_D = \partial/\partial y^D$$
$$D_k = \partial/\partial y_r{}^k + i\, \partial/\partial y_i{}^k$$

(10.64)

for k = 1, 2, … , (D – 1). The action

$$I = \int d^{(D-1)}y \mathcal{L}_u \tag{10.65}$$

It is easy to show that this action is also real.

10.3.2.4 Type IV "Down-Quark-like" Tachyon Universe Particle Lagrangian

The lagrangian density of a free "down-quark-like" universe particle is

$$\mathcal{L}_d = \psi_d^{\ C}(y)(\gamma^{\mathbf{D}}\partial/\partial t + \boldsymbol{\gamma}\cdot(\nabla_{\mathbf{r}} + i\nabla_{\mathbf{i}}) - m)\psi_d(y) \tag{10.66}$$

where

$$\psi_d^{\ C}(y) = [\psi_d(y)]^\dagger|_{\mathbf{y_i} = -\mathbf{y_i}}\, i\gamma^D\gamma^5 \tag{10.67}$$

In words, eq. 10.67 states: take the hermitean conjugate of $\psi_d(y)$; change $\mathbf{y}_i$ to $-\mathbf{y}_i$; and then post-multiply by the indicated factors.

The action is

$$I = \int d^{(D-1)}y \mathcal{L}_d \tag{10.68}$$

The action is real. The lagrangian can also be separated into left-handed and right-handed parts using projection operators.

10.3.3 Form of The Megaverse Quantum Coordinates Gauge Field

The discussions of sections 10.3.1 and 10.3.2 assumed the coordinates were Megaverse coordinates and their derivatives. Prior to those discussions we indicated we would use quantum coordinates in the Megaverse of the form[133]

$$Y^i(y) = y^i + i\, Y_u^{\ i}(y)/M_u^{\ D/2} \tag{10.4}$$

and their derivatives

$$\partial_i = \partial/\partial Y^i(y) = \partial/\partial(y^i - Y_u^{\ i}(y)/M_u^{\ D/2}) \tag{10.5}$$

for i = 1, 2, … , D to eliminate divergences in quantum field theory. The subscript "u" signifies universes. The mass constant for the Megaverse, M_u, may be the same as the mass constant M_c appearing in the Two Tier mechanism for our universe. (See chapter 7 for a discussion of eliminating infinities with this mechanism.)

In this section we define the gauge fields $Y_u^{\ i}(y)$ of the Megaverse.[134] They are similar to the $Y^\mu(y)$ fields of our New Standard Model.[135] The $Y_u(y)$ D-dimensional

---

[133] The denominator $M_u^{\ D/2}$ is necessitated by the dimension of $Y_u^{\ i}(y)$ which is $[m]^{D/2-1}$. Eqs. 10.78 and 10.81 below imply this conclusion.

[134] This choice implies that the Megaverse Y mass $M_u = M_C$, its universe mass.

[135] See Blaha (2005a) for details.

vector gauge field, in the absence of external sources, will be defined in a D-dimensional Coulomb gauge:

$$Y_u^{\,D}(y) = 0 \qquad (10.69)$$
$$\partial Y_u^{\,j}(y)/\partial y^j = 0$$

where the sum over j is over the D – 1 spatial y coordinates. We follow a procedure similar to Blaha (2003) but for D-dimensional space. The lagrangian density for the free $Y_u^{\,j}(y)$ fields is

$$\mathscr{L}_u = -\tfrac{1}{4}\, F_u^{\,\mu\nu} F_{u\mu\nu} \qquad (10.70)$$

and the lagrangian is

$$L_u = \int d^{(D-1)}y\, \mathscr{L}_u \qquad (10.71a)$$

with

$$F_{u\mu\nu} = \partial Y_{u\mu}/\partial y^\nu - \partial Y_{u\nu}/\partial y^\mu \qquad (10.71b)$$

The equal time commutation relations, derived in the usual way, are:

$$[Y_u^{\,\mu}(\mathbf{y}, y^0), Y_u^{\,\nu}(\mathbf{y}', y^0)] = [\pi_u^{\,\mu}(\mathbf{y}, y^0), \pi_u^{\,\nu}(\mathbf{y}', y^0)] = 0 \qquad (10.72)$$
$$[\pi_u^{\,j}(\mathbf{y}, y^0), Y_{uk}(\mathbf{y}', y^0)] = -i\, \delta^{(D-1)tr}{}_{jk}(\mathbf{y} - \mathbf{y}') \qquad (10.73)$$

for μ, ν, j, k = 1, 2, … , (D – 1) where

$$\pi_u^{\,k} = \partial \mathscr{L}_u/\partial Y_{uk}' \qquad (10.74)$$
$$\pi_u^{\,0} = 0 \qquad (10.75)$$

and

$$\delta^{tr}{}_{jk}(\mathbf{y} - \mathbf{y}') = \int d^{(D-1)}k\, e^{i\,\mathbf{k}\bullet(\mathbf{y}-\mathbf{y}')}\, (\delta_{jk} - k_j k_k/\mathbf{k}^2)/(2\pi)^{D-1} \qquad (10.76)$$

$$Y_{uk}' = \partial Y_{uk}/\partial y^D \qquad (10.77)$$

The Coulomb gauge indicates D – 2 degrees of freedom are present in the vector potential. The Fourier expansion of the vector potential is:

$$Y_u^{\,i}(y) = \int d^{(D-1)}k\, N_0(k) \sum_{\lambda=1}^{D-2} \varepsilon^i(k, \lambda)[a(k,\lambda)\, e^{-ik\cdot y} + a^\dagger(k,\lambda)\, e^{ik\cdot y}] \qquad (10.78)$$

where

$$N_0(k) = [(2\pi)^{(D-1)} 2\omega_k]^{-1/2} \qquad (10.79)$$

and (since the field is massless)

$$k^D = \omega_k = (\mathbf{k}^2)^{1/2} \qquad (10.80)$$

where k^D is the energy, and where the $\varepsilon^i(k, \lambda)$ are the polarization unit vectors for $\lambda = 1, \ldots, (D-2)$ and $k^\mu k_\mu = k^{D\,2} - \mathbf{k}^2 = 0$.

The commutation relations of the Fourier coefficient operators are:

$$[a(k,\lambda), a^\dagger(k',\lambda')] = \delta_{\lambda\lambda'}\delta^{(D-1)}(\mathbf{k} - \mathbf{k'}) \tag{10.81}$$
$$[a^\dagger(k,\lambda), a^\dagger(k',\lambda')] = [a(k,\lambda), a(k',\lambda')] = 0 \tag{10.82}$$

and the polarization vectors satisfy

$$\sum_{\lambda=1}^{D-2} \varepsilon_i(k, \lambda)\varepsilon_j(k, \lambda) = (\delta_{ij} - k_ik_j/\mathbf{k}^2) \tag{10.83}$$

It will be convenient to divide the Y field into positive and negative frequency parts:

$$Y_u{}^+{}_i(y) = \int d^{(D-1)}k\, N_0(k) \sum_{\lambda=1}^{D-2} \varepsilon_i(k, \lambda)\, a(k,\lambda)\, e^{-ik\cdot y} \tag{10.84}$$

and

$$Y_u{}^-{}_i(y) = \int d^{(D-1)}k\, N_0(k) \sum_{\lambda=1}^{D-2} \varepsilon_i(k, \lambda)\, a^\dagger(k,\lambda)\, e^{ik\cdot y} \tag{10.85}$$

For later use we note the commutator between the positive and negative frequency parts is:

$$[\, Y_u{}^-{}_j(y_1), Y_u{}^+{}_k(y_2)] = -\int d^{(D-1)}k\, e^{ik\cdot(y_1 - y_2)}\, (\delta_{jk} - k_jk_k/\mathbf{k}^2)/[(2\pi)^{D-1}\, 2\omega_k] \tag{10.86}$$

10.3.3.1 Y^μ Fock Space Imaginary Coordinate States

States can also be defines for the quantized Y^μ field. These states will be similar in form to electromagnetic photon states but play a different role in our approach since they are in fact coordinate excitation states for the imaginary part of $Y^i(y)$ (eq. 10.4). Thus universe particles (and other fields) will exist in a real D-dimensional space with quantum excitations into imaginary Quantum Dimensions. These excitations become significant at high energies. At low energies space appears as c-number complex; at very high energies space becomes slightly q-number complex.

There are two types of imaginary coordinate excitations: 1.) Quantum excitations into Fock states consisting of a superposition of states with a definite finite number of Y_u "particles" and 2.) Imaginary coordinate excitations into coherent Y_u states with an "infinite" number of particles. Coherent states can be viewed as representing "classical" fields.

In this section we will consider Y_u field states with a definite number of excitations ("particles"). The raising and lowering operators of the Y_u field can be used to define free particle states. For example a one particle state can be defined by

$$|k, \lambda> = a^{\dagger}(k, \lambda)|0> \tag{10.87}$$

with corresponding bra state

$$<k, \lambda| = <0|a(k, \lambda) \tag{10.88}$$

where the "coordinate vacuum" is defined as usual:

$$a(k, \lambda)|0> = 0 \tag{10.89}$$

$$<0|a^{\dagger}(k, \lambda) = 0 \tag{10.90}$$

Multi-particle states can also be defined in the conventional way with products of the raising and lowering operators applied to the vacuum. The set of all states containing a finite number of "particles" constitutes a Fock space.

A state with a finite number of Y_u "particles" represents a quantum fluctuation into imaginary Quantum Dimensions.

10.3.3.2 Y_u Coherent Imaginary Coordinate States

Coherent Y_u states bring us closer what we might consider to be "classical" imaginary dimensions – dimensions that we can, in principle, experience as we do normal dimensions. Let us define the coherent state[136]

$$| y, p> = e^{-\mathbf{p}\cdot\mathbf{Y}_u^-(y)/M_u^{D/2}}|0> \tag{10.91}$$

This state is an eigenstate of the coordinate operator $Y_u^+(y')$:

$$Y_u^+{}_j(y_1)\,|y_2, p> = -[Y_u^+{}_j(y_1), \mathbf{p}\cdot\mathbf{Y}^-(y_2)]/M_u^{D/2}|y, p> \tag{10.92}$$

$$= -\int d^{D-1}k\, [N_0(k)]^2\, e^{ik\cdot(y_2 - y_1)}\, (p_j - k_j\mathbf{p}\cdot\mathbf{k}/\mathbf{k}^2)/M_u^{D/2}|y, p>$$
$$= p^i\Delta_{Tij}(y_1 - y_2)/M_u^{D/2}|y, p> \tag{10.93}$$

where $p^i\Delta_{Tij}(y_1 - y_2)/M_u^{D/2}$ is the eigenvalue of $Y_u^+{}_j(y_1)$. As we will see later, the eigenvalue of Y_u^+ becomes large as $(y_1 - y_2)^2 \rightarrow 0$. Thus the imaginary Quantum Dimensions become significant at very short distances, and then significantly modifies the high-energy behavior of quantum field theories. In particular, Quantum Dimensions have a significant effect when

$$(y_1 - y_2)^2 \lessapprox (2^{D-2}\pi^{D-2}M_u^2)^{-1} \tag{10.94}$$

---

[136] Coherent states are well known in the physics literature. See for example T. W. B. Kibble, J. Math. Phys. **9**, 315 (1968) and references therein; V. Chung, Phys. Rev. **140**, B1110 (1965); J. R. Klauder, J. McKenna, and E. J. Woods, J. Math. Phys. **7**, 822 (1966) and references therein.

We assume the mass scale $M_u = M_C$ is very large – perhaps of the order of the Planck mass (1.221×10^{19} GeV/c^2).

10.3.3.3 Quantization of the Type I Free Universe Particle Dirac Field

The quantization procedure is formally identical to that of a conventional Dirac particle. The standard equal time anti-commutation relations for a D-dimensional fermion field are:

$$\{\psi_\alpha(Y), \psi_\beta(Y')\} = \{\pi_{\psi\alpha}(Y), \pi_{\psi\beta}(Y')\} = 0 \qquad (10.95)$$
$$\{\pi_{\psi\alpha}(Y), \psi_\beta(Y')\} = i\,\delta_{\alpha\beta}\,\delta^{D-1}(\mathbf{Y} - \mathbf{Y'}) \qquad (10.96)$$

where α and β are the spinor indices ranging from 1 to $N_{MRC} = 2^{D/2}$ and where

$$\pi_{\psi\alpha}(Y) = i\,\psi_\alpha^\dagger(Y) \qquad (10.97)$$

The field can be expanded in a fourier series:

$$\psi(Y(y)) = \sum_s \int d^{D-1}p\, N^d{}_m(p)\, [b(p,s)u(p,s)\, :e^{-ip\cdot(y + iYu/M_u{}^{D/2})}: + d^\dagger(p,s)v(p,s)\, :\exp(ip\cdot(y + iYu/Mu^{D/2})):] \qquad (10.98)$$

$$\psi^\dagger(Y(y)) = \sum_s \int d^{D-1}p\, N^d{}_m(p)\, [b^\dagger(p,s)\bar{u}(p,s)\gamma^0\, :e^{+ip\cdot(y + iYu/M_u{}^{D/2})}: + d(p,s)\bar{v}(p,s)\gamma^0 \cdot :\exp(-ip\cdot(y + iYu/Mu^{D/2}):] \qquad (10.99)$$

where

$$N^d{}_m(p) = [m/((2\pi)^{D-1}E_p)]^{1/2} \qquad (10.100)$$

and

$$E_p = p^D = (\mathbf{p}^2 + m^2)^{1/2} \qquad (10.101)$$

with : ... : signifying normal ordering. The commutation relations of the Fourier coefficient operators are:

$$\{b(p,s), b^\dagger(p',s')\} = \delta_{ss'}\,\delta^{D-1}(\mathbf{p} - \mathbf{p'}) \qquad (10.102)$$
$$\{d(p,s), d^\dagger(p',s')\} = \delta_{ss'}\,\delta^{D-1}(\mathbf{p} - \mathbf{p'}) \qquad (10.103)$$
$$\{b(p,s), b(p',s')\} = \{d(p,s), d(p',s')\} = 0 \qquad (10.104)$$
$$\{b^\dagger(p,s), b^\dagger(p',s')\} = \{d^\dagger(p,s), d^\dagger(p',s')\} = 0 \qquad (10.105)$$
$$\{b(p,s), d^\dagger(p',s')\} = \{d(p,s), b^\dagger(p',s')\} = 0 \qquad (10.106)$$
$$\{b^\dagger(p,s), d^\dagger(p',s')\} = \{d(p,s), b(p',s')\} = 0 \qquad (10.107)$$

The spinors u(p,s) and v(p,s) are defined in a conventional way (as in Bjorken and Drell). However their form is different from the 4-dimensional case. If one takes the $N_{MRC} \times N_{MRC} \equiv 2^{D/2} \times 2^{D/2}$ $\gamma\cdot p$ matrix, then the first $2^{D/2-1}$ columns give u(p,s) up to a normalization for the free particle case, the remaining $2^{D/2-1}$ columns give v(p,s) up to a normalization.

10.3.3.4 Feynman Propagators for the Type I Free Universe Particle Dirac Field

The form of the fermionic universe particle Feynman propagator differs from a conventional fermion propagator by having a Gaussian factor R(**p**, z) in its fourier expansions. This follows from using quantum Megaverse coordinates (eq. 10.4).

$$iS_F{}^{TT}(y_1 - y_2) = \langle 0|T(\bar{\psi}(Y(y_1))\psi(Y(y_2)))|0\rangle \tag{10.108}$$

where the time ordering is with respect to $y_1{}^D$ and $y_2{}^D$. Expanding the free fields leads to the fourier representation:

$$iS_F{}^{TT}(y_1 - y_2) = i\int \frac{d^Dp\, e^{-ip\cdot(y_1 - y_2)}\,(\not{p}+ m)\, R(\mathbf{p}, y_1 - y_2)}{(2\pi)^D\,(p^2 - m^2 + i\varepsilon)} \tag{10.109}$$

where

$$R(\mathbf{p}, y_1 - y_2) = \exp[-p^ip^j\Delta_{Tij}(y_1 - y_2)/M_u{}^D] \tag{10.110}$$
$$= \exp\{-p^2[A(v) + B(v)\cos^2\theta] / [(2\pi)^{D-2}M_c{}^4z^2]\} \tag{10.111}$$

(Note p^2 is the square of the spatial (D – 1)-vector.) with

$$z^\mu = y_1{}^\mu - y_2{}^\mu \tag{10.112}$$
$$z = |\mathbf{z}| = |\mathbf{y_1} - \mathbf{y_2}| \tag{10.113}$$
$$p = |\mathbf{p}| \tag{10.114}$$
$$v = |z^0|/z \tag{10.115}$$
$$A(v) = (1 - v^2)^{-1} + .5v\ln[(v - 1)/(v + 1)] \tag{10.116}$$
$$B(v) = v^2(1 - v^2)^{-1} - 1.5v\ln[(v - 1)/(v + 1)] \tag{10.117}$$
$$\mathbf{p}\cdot\mathbf{z} = pz\cos\theta \tag{10.118}$$

and $|\mathbf{p}|$ denoting the length of a spatial (D – 1)-vector **p** while $|z^0|$ is the absolute value of $z^0 \equiv z^D$.

As eq. 10.109 indicates, the Gaussian damping factor[137] R(p, z) for large spatial momentum p is the same for both the positive and negative frequency parts of the Two Tier Feynman propagator. We are assuming the spatial momentum is real-valued in this discussion. It is also important to note that R(p, z) does not depend on $p^0 = p^D$ (in the Y Coulomb gauge) and thus the integration over p^0 proceeds in the usual way to produce time-ordered positive and negative frequency parts.

---

[137] Note the Gaussian damping is for all D – 1 spatial momentum integrations.

10.3.3.5 Feynman Propagators for the Types II, III, and IV Free Universe Particle Dirac Fields

These propagators differ in details from the Type I propagator. The differences modulo the change in dimension appear in Blaha (2011c). See also Blaha (2005a) for a detailed discussion of 4-dimensional spin ½ particle propagators.

10.3.4 Expanding and Contracting Universes: Impact of Time Dependent Universe Particle Masses

Our discussions of the dynamics of universe particles assumed their masses were constant. However the definition of mass in terms of the area of a universe based on the physics of black holes is

$$M = \kappa A/8\pi \tag{10.119}$$

where A is the area of the black hole shows that *the mass of a universe particle is time dependent* because the area of a universe is generally time dependent. For example, our universe is expanding and its surface area is thus growing with time.

Eqs. 10.11 (and subsequent fermionic dynamic equations) must then be modified from

$$\psi(y) = \exp[-imt]w(0) \tag{10.11}$$

to a covariant form:

$$\psi(y) = \exp[-i\int_0^{w\cdot y} m(t')dt']w(0) \tag{10.120}$$

where w is a unit D-vector in the time (y^D) direction ($w^2 = 1$). The lower bound on the integral, 0, is the time of the beginning of the universe particle – its Big Bang. Thus the cumulative change in the mass of the universe particle may be significant. It is interesting to note that the Wheeler-Dewitt equation also has a variable value mass term R that also depends on the evolution of the universe.

Eq. 10.120 satisfies the free covariant Dirac-like universe particle field dynamic equation

$$[i\gamma^i\partial/\partial y^i - m(w\cdot y)]\psi(y) = 0 \tag{10.121}$$

In contrast to the constant mass equation eq. 10.19. Substituting eq. 10.120 in eq. 10.121 we find

$$(\gamma^i w_i\, m(w\cdot y) - m(w\cdot y))\psi(y) = 0 \tag{10.122}$$

or

$$(\gamma^i w_i - 1)\psi(y) = 0 \tag{10.123}$$

Upon performing a D-dimensional Lorentz boost (of the type of eqs. 10.13 – 10.16) on eq. 10.123 we obtain

$$(\gamma_i p^i/m_0 - 1)\psi(y) = 0$$

or

$$(\gamma_i p^i - m_0)\psi(y) = 0 \qquad (10.124)$$

where p^i is a momentum D-vector with $p^2 = m_0^2$. Eq. 10.123 is the constant mass momentum space dynamic equation. It determines the spinor in $\psi(y)$. After taking account of the quantum coordinates the quantum Dirac-like universe particle wave function has the form

$$\psi(Y(y)) = \sum_s \int d^{(D-1)}p\, N^d{}_m(p)\, [b(p,s)u(p,s) : \exp[-iG(p, Y(y))]: + d^\dagger(p,s)v(p,s) \cdot \cdot :\exp[+iG(p, Y(y))]:\} \qquad (10.125)$$

$$\psi^\dagger(Y(y)) = \sum_s \int d^{(D-1)}p\, N^d{}_m(p)\, \{b^\dagger(p,s)\bar{u}(p,s)\gamma^0 :\exp[+iG(p, Y(y))]: + d(p,s)\bar{v}(p,s)\gamma^0 \cdot \cdot \exp[-iG(p, Y(y))]:\} \qquad (10.126)$$

where : … : denotes normal ordering and

$$G(p, Y(y)) = \int_0^{p\cdot Y(y)/\lambda} m(t')dt' \qquad (10.127)$$

with $\lambda = m_0$, and $N^d{}_m(p)$ a normalization constant. Contrast eqs. 10.125-10.126 to the constant mass case eqs. 10.98-10.101. The *constant mass case* simply sets $m(t') = m_0$.

If we examine the integral eq. 10.127 for a short time interval δt in the particle's rest frame then $G(p, Y(y)) \approx m(0)\delta t$ and so we define $m(0) = m_0$. Based on the formula for universe particle mass (eq. 10.119) we anticipate that m_0 might be as large as the Planck mass or larger – thus an extremely short radius. Blaha (2013) describes a quantum Big Bang model in which the initial radius of the universe is $O(EM_{Planck}^{-2})$ where E is of the order of 1 and has the dimensions of [mass].

Thus we have a closed form definition of a quantum universe particle wave function for universe particles of type I. A similar procedure can be followed for universe particles of types II, III, and IV.

The Feynman propagator for type I quantum fields is *not* eq. 10.109 but now has a form reflecting the Y(y) dependence of the quantum fields in eqs. 10.125 and 10.126:

$$iS_F{}^{TT}(y_1, y_2) = i \int \frac{d^Dp\, \{ \langle 0|\theta(y_{1D} - y_{2D})G(y_1, y_2) + \theta(y_{2D} - y_{1D})G(y_2, y_1)\}0\rangle}{(2\pi)^D\, (p\!\!\!/ - m_0)} \qquad (10.128)$$

where p^D is the energy and

$$G(y_1, y_2) = :\exp[-iG(p, Y(y_1))]: :\exp[+iG(p, Y(y_2))]: \qquad (10.129)$$

Let

$$G_{tot}(y_1, y_2) = \langle 0|\theta(y_{1D} - y_{2D})G(y_1, y_2) + \theta(y_{2D} - y_{1D})G(y_2, y_1)\}0\rangle \qquad (10.130)$$
$$= \langle 0|\theta(y_{1D} - y_{2D}):\exp[-iG(p, Y(y_1))]::\exp[+iG(p, Y(y_2))]: +$$
$$+ \theta(y_{2D} - y_{1D}) :\exp[-iG(p, Y(y_2))]::\exp[+iG(p, Y(y_1)):]|0\rangle$$

$$= \langle 0|\theta(y^D{}_1 - y^D{}_2): \exp[-i\int_0^{p\cdot Y(y1)/\lambda} m(t')dt']::\exp[+i\int_0^{p\cdot Y(y2)/\lambda} m(t')dt']: +$$

$$+ \theta(y^D{}_2 - y^D{}_1):\exp[-i \exp[+i\int_0^{p\cdot Y(y2)/\lambda} m(t')dt']::\exp[+i\int_0^{p\cdot Y(y1)/\lambda} m(t')dt']:|0\rangle$$

with $\lambda = m_0$ then

$$iS_F{}^{TT}(y_1, y_2) = i\int \frac{d^Dp\ G_{tot}(y_1, y_2)}{(2\pi)^D (p - m_0)} \qquad (10.131)$$

Except for the case of a constant mass, where $m(t) = m_0$, the Feynman propagator is not a function of $y_1 - y_2$. The evaluation of eq. 10.130 in the general case of a variable mass is straightforward but cumbersome. For the special case of a linear time dependence of the mass, $m(t) = at$, we find eq. 10.130 gives

$$G_{tot}(y_1, y_2) = \langle 0|\theta(y^D{}_1 - y^D{}_2):\exp[-ia(p\cdot Y(y_1)/m_0)^2/2]::\exp[+ia(p\cdot Y(y_2)/m_0)^2/2]: +$$
$$+ \theta(y^D{}_1 - y^D{}_2):\exp[-ia(p\cdot Y(y_2)/m_0)^2/2]::\exp[+ia(p\cdot Y(y_1)/m_0)^2/2]:|0\rangle \qquad (10.132)$$

yielding a complex function of p, y_1, and y_2. *Note that the lower bound of the integrals in the Feynman propagator cancel and thus the need for an understanding of the beginning of a universe is removed in this case.*

We have shown that universe particle theory can handle the case of a variable universe mass m(t). Expanding or contracting (or oscillating) universe particles correspond to expanding and contracting (or oscillating) universes.

REFERENCES

Akhiezer, N. I., Frink, A. H. (tr), 1962, *The Calculus of Variations* (Blaisdell Publishing, New York, 1962).

Bjorken, J. D., Drell, S. D., 1964, *Relativistic Quantum Mechanics* (McGraw-Hill, New York, 1965).

Bjorken, J. D., Drell, S. D., 1965, *Relativistic Quantum Fields* (McGraw-Hill, New York, 1965).

Blaha, S., 1998, *Cosmos and Consciousness* (Pingree-Hill Publishing, Auburn, NH, 1998).

_______, 2002, *A Finite Unified Quantum Field Theory of the Elementary Particle Standard Model and Quantum Gravity Based on New Quantum Dimensions™ & a New Paradigm in the Calculus of Variations* (Pingree-Hill Publishing, Auburn, NH, 2002).

_______, 2003, *A Finite Unified Quantum Field Theory of the Elementary Particle Standard Model and Quantum Gravity Based on New Quantum Dimensions™ and a New Paradigm in the Calculus of Variations* (Pingree-Hill Publishing, Auburn, NH, 2003).

_______, 2004, *Quantum Big Bang Cosmology: Complex Space-time General Relativity, Quantum Coordinates™ Dodecahedral Universe, Inflation, and New Spin 0, ½, 1 & 2 Tachyons & Imagyons* (Pingree-Hill Publishing, Auburn, NH, 2004).

______, *2005a, Quantum Theory of the Third Kind: A New Type of Divergence-free Quantum Field Theory Supporting a Unified Standard Model of Elementary Particles and Quantum Gravity based on a New Method in the Calculus of Variations* (Pingree-Hill Publishing, Auburn, NH, 2005).

______, 2005b, *The Metatheory of Physics Theories, and the Theory of Everything as a Quantum Computer Language* (Pingree-Hill Publishing, Auburn, NH, 2005).

______, 2005c, *The Equivalence of Elementary Particle Theories and Computer Languages: Quantum Computers, Turing Machines, Standard Model, Superstring Theory, and a Proof that Gödel's Theorem Implies Nature Must Be Quantum* (Pingree-Hill Publishing, Auburn, NH, 2005).

______, 2006a, *The Foundation of the Forces of Nature* (Pingree-Hill Publishing, Auburn, NH, 2006).

______, 2006b, *A Derivation of ElectroWeak Theory based on an Extension of Special Relativity; Black Hole Tachyons; & Tachyons of Any Spin.* (Pingree-Hill Publishing, Auburn, NH, 2006).

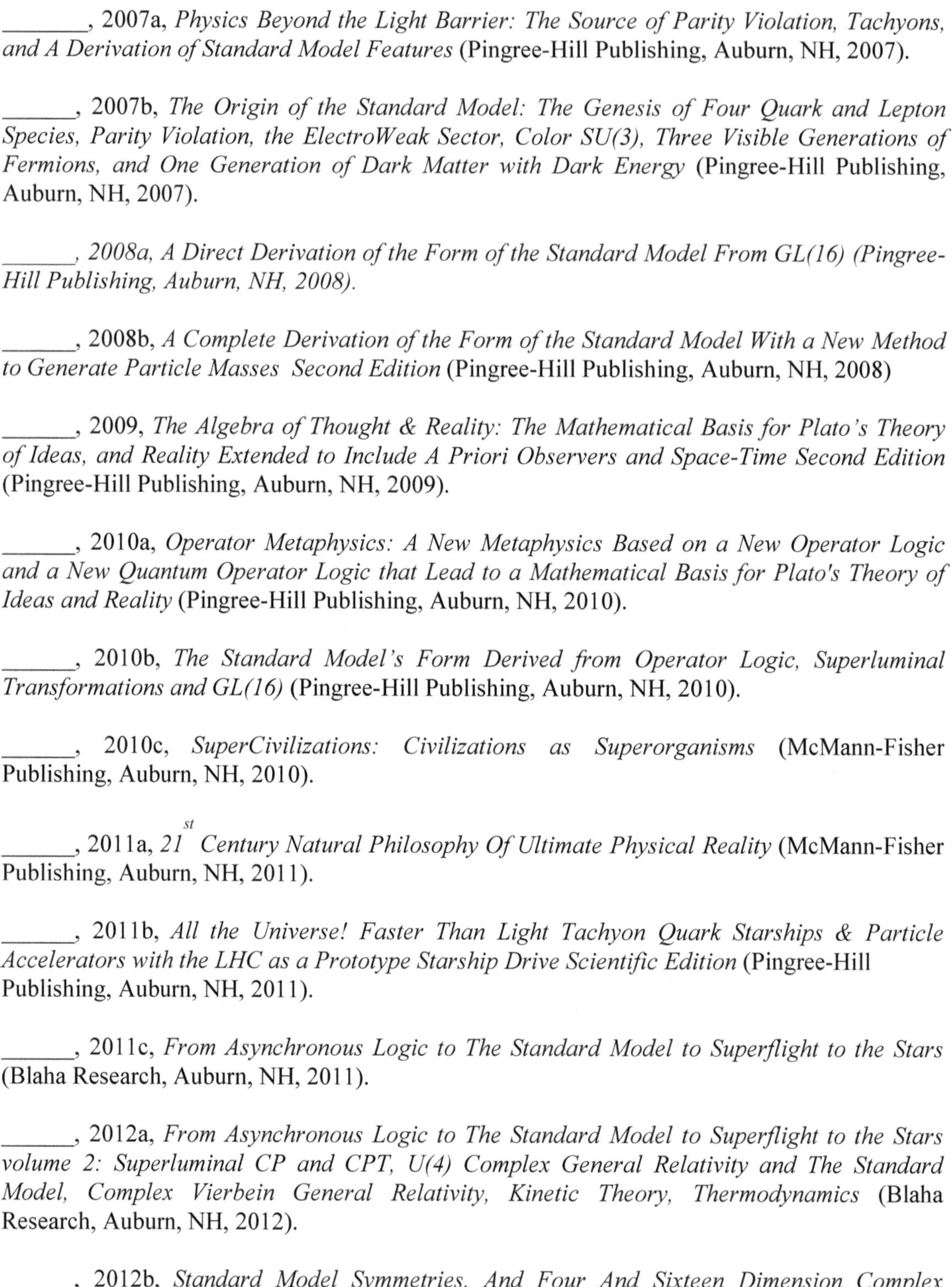

_______, 2007a, *Physics Beyond the Light Barrier: The Source of Parity Violation, Tachyons, and A Derivation of Standard Model Features* (Pingree-Hill Publishing, Auburn, NH, 2007).

______, 2007b, *The Origin of the Standard Model: The Genesis of Four Quark and Lepton Species, Parity Violation, the ElectroWeak Sector, Color SU(3), Three Visible Generations of Fermions, and One Generation of Dark Matter with Dark Energy* (Pingree-Hill Publishing, Auburn, NH, 2007).

______, *2008a, A Direct Derivation of the Form of the Standard Model From GL(16) (Pingree-Hill Publishing, Auburn, NH, 2008).*

______, 2008b, *A Complete Derivation of the Form of the Standard Model With a New Method to Generate Particle Masses Second Edition* (Pingree-Hill Publishing, Auburn, NH, 2008)

______, 2009, *The Algebra of Thought & Reality: The Mathematical Basis for Plato's Theory of Ideas, and Reality Extended to Include A Priori Observers and Space-Time Second Edition* (Pingree-Hill Publishing, Auburn, NH, 2009).

______, 2010a, *Operator Metaphysics: A New Metaphysics Based on a New Operator Logic and a New Quantum Operator Logic that Lead to a Mathematical Basis for Plato's Theory of Ideas and Reality* (Pingree-Hill Publishing, Auburn, NH, 2010).

______, 2010b, *The Standard Model's Form Derived from Operator Logic, Superluminal Transformations and GL(16)* (Pingree-Hill Publishing, Auburn, NH, 2010).

______, 2010c, *SuperCivilizations: Civilizations as Superorganisms* (McMann-Fisher Publishing, Auburn, NH, 2010).

______, 2011a, *21st Century Natural Philosophy Of Ultimate Physical Reality* (McMann-Fisher Publishing, Auburn, NH, 2011).

______, 2011b, *All the Universe! Faster Than Light Tachyon Quark Starships & Particle Accelerators with the LHC as a Prototype Starship Drive Scientific Edition* (Pingree-Hill Publishing, Auburn, NH, 2011).

______, 2011c, *From Asynchronous Logic to The Standard Model to Superflight to the Stars* (Blaha Research, Auburn, NH, 2011).

______, 2012a, *From Asynchronous Logic to The Standard Model to Superflight to the Stars volume 2: Superluminal CP and CPT, U(4) Complex General Relativity and The Standard Model, Complex Vierbein General Relativity, Kinetic Theory, Thermodynamics* (Blaha Research, Auburn, NH, 2012).

______, 2012b, *Standard Model Symmetries, And Four And Sixteen Dimension Complex Relativity; The Origin Of Higgs Mass Terms* (Blaha Reasearch, Auburn, NH, 2012).

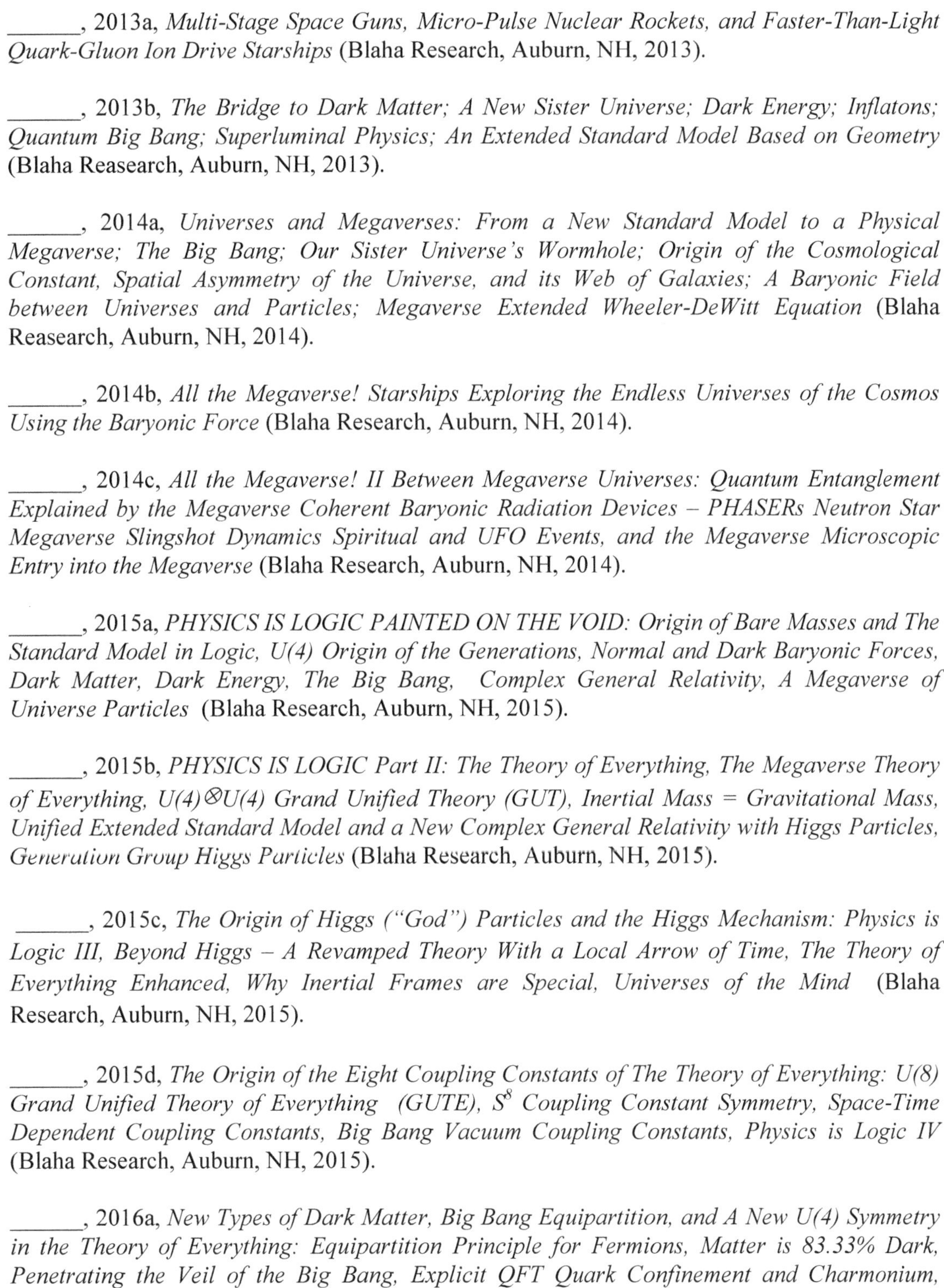

______, 2013a, *Multi-Stage Space Guns, Micro-Pulse Nuclear Rockets, and Faster-Than-Light Quark-Gluon Ion Drive Starships* (Blaha Research, Auburn, NH, 2013).

______, 2013b, *The Bridge to Dark Matter; A New Sister Universe; Dark Energy; Inflatons; Quantum Big Bang; Superluminal Physics; An Extended Standard Model Based on Geometry* (Blaha Reasearch, Auburn, NH, 2013).

______, 2014a, *Universes and Megaverses: From a New Standard Model to a Physical Megaverse; The Big Bang; Our Sister Universe's Wormhole; Origin of the Cosmological Constant, Spatial Asymmetry of the Universe, and its Web of Galaxies; A Baryonic Field between Universes and Particles; Megaverse Extended Wheeler-DeWitt Equation* (Blaha Reasearch, Auburn, NH, 2014).

______, 2014b, *All the Megaverse! Starships Exploring the Endless Universes of the Cosmos Using the Baryonic Force* (Blaha Research, Auburn, NH, 2014).

______, 2014c, *All the Megaverse! II Between Megaverse Universes: Quantum Entanglement Explained by the Megaverse Coherent Baryonic Radiation Devices – PHASERs Neutron Star Megaverse Slingshot Dynamics Spiritual and UFO Events, and the Megaverse Microscopic Entry into the Megaverse* (Blaha Research, Auburn, NH, 2014).

______, 2015a, *PHYSICS IS LOGIC PAINTED ON THE VOID: Origin of Bare Masses and The Standard Model in Logic, U(4) Origin of the Generations, Normal and Dark Baryonic Forces, Dark Matter, Dark Energy, The Big Bang, Complex General Relativity, A Megaverse of Universe Particles* (Blaha Research, Auburn, NH, 2015).

______, 2015b, *PHYSICS IS LOGIC Part II: The Theory of Everything, The Megaverse Theory of Everything, U(4)⊗U(4) Grand Unified Theory (GUT), Inertial Mass = Gravitational Mass, Unified Extended Standard Model and a New Complex General Relativity with Higgs Particles, Generation Group Higgs Particles* (Blaha Research, Auburn, NH, 2015).

______, 2015c, *The Origin of Higgs ("God") Particles and the Higgs Mechanism: Physics is Logic III, Beyond Higgs – A Revamped Theory With a Local Arrow of Time, The Theory of Everything Enhanced, Why Inertial Frames are Special, Universes of the Mind* (Blaha Research, Auburn, NH, 2015).

______, 2015d, *The Origin of the Eight Coupling Constants of The Theory of Everything: U(8) Grand Unified Theory of Everything (GUTE), S^8 Coupling Constant Symmetry, Space-Time Dependent Coupling Constants, Big Bang Vacuum Coupling Constants, Physics is Logic IV* (Blaha Research, Auburn, NH, 2015).

______, 2016a, *New Types of Dark Matter, Big Bang Equipartition, and A New U(4) Symmetry in the Theory of Everything: Equipartition Principle for Fermions, Matter is 83.33% Dark, Penetrating the Veil of the Big Bang, Explicit QFT Quark Confinement and Charmonium, Physics is Logic V* (Blaha Research, Auburn, NH, 2016).

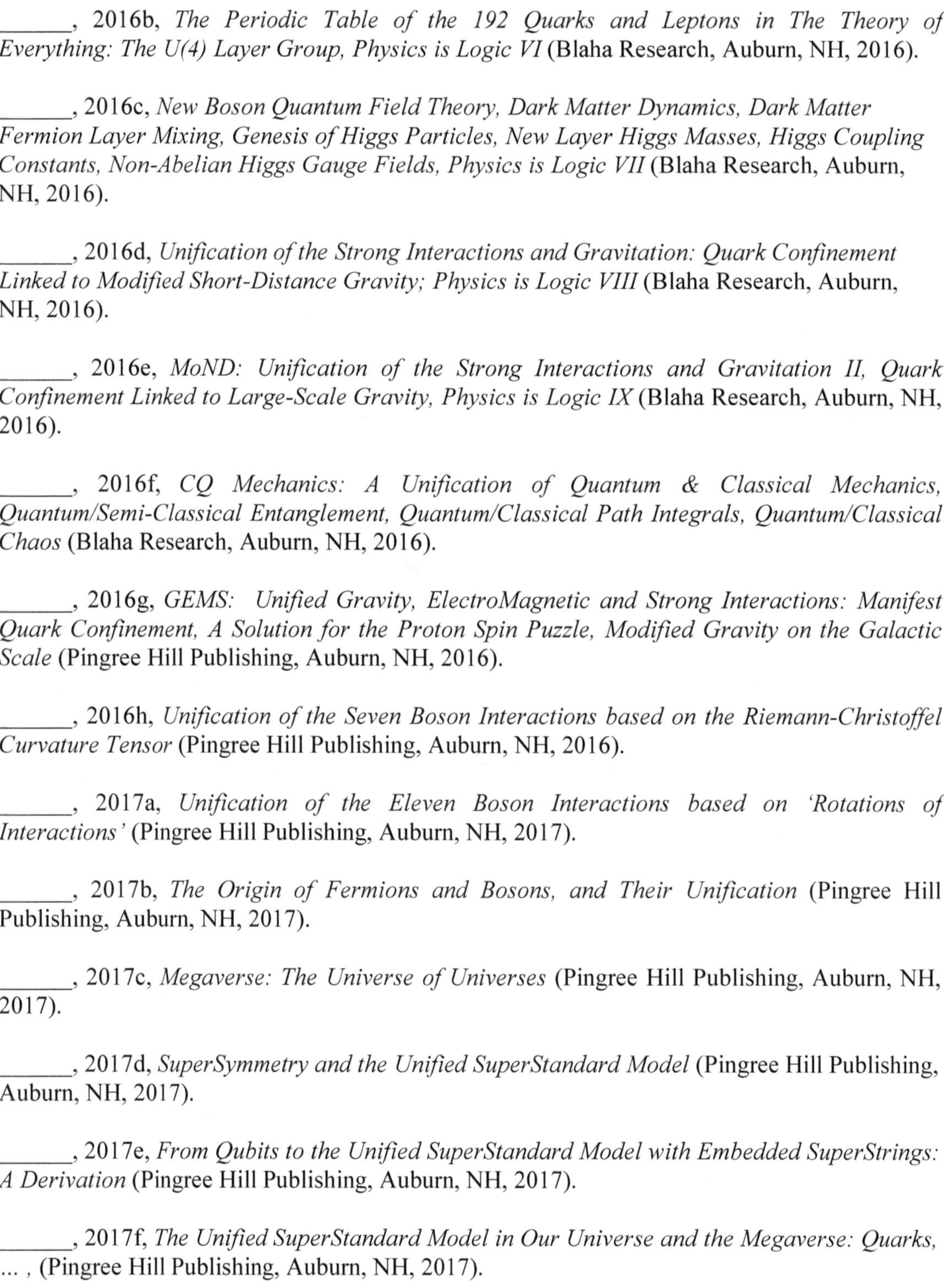

______, 2016b, *The Periodic Table of the 192 Quarks and Leptons in The Theory of Everything: The U(4) Layer Group, Physics is Logic VI* (Blaha Research, Auburn, NH, 2016).

______, 2016c, *New Boson Quantum Field Theory, Dark Matter Dynamics, Dark Matter Fermion Layer Mixing, Genesis of Higgs Particles, New Layer Higgs Masses, Higgs Coupling Constants, Non-Abelian Higgs Gauge Fields, Physics is Logic VII* (Blaha Research, Auburn, NH, 2016).

______, 2016d, *Unification of the Strong Interactions and Gravitation: Quark Confinement Linked to Modified Short-Distance Gravity; Physics is Logic VIII* (Blaha Research, Auburn, NH, 2016).

______, 2016e, *MoND: Unification of the Strong Interactions and Gravitation II, Quark Confinement Linked to Large-Scale Gravity, Physics is Logic IX* (Blaha Research, Auburn, NH, 2016).

______, 2016f, *CQ Mechanics: A Unification of Quantum & Classical Mechanics, Quantum/Semi-Classical Entanglement, Quantum/Classical Path Integrals, Quantum/Classical Chaos* (Blaha Research, Auburn, NH, 2016).

______, 2016g, *GEMS: Unified Gravity, ElectroMagnetic and Strong Interactions: Manifest Quark Confinement, A Solution for the Proton Spin Puzzle, Modified Gravity on the Galactic Scale* (Pingree Hill Publishing, Auburn, NH, 2016).

______, 2016h, *Unification of the Seven Boson Interactions based on the Riemann-Christoffel Curvature Tensor* (Pingree Hill Publishing, Auburn, NH, 2016).

______, 2017a, *Unification of the Eleven Boson Interactions based on 'Rotations of Interactions'* (Pingree Hill Publishing, Auburn, NH, 2017).

______, 2017b, *The Origin of Fermions and Bosons, and Their Unification* (Pingree Hill Publishing, Auburn, NH, 2017).

______, 2017c, *Megaverse: The Universe of Universes* (Pingree Hill Publishing, Auburn, NH, 2017).

______, 2017d, *SuperSymmetry and the Unified SuperStandard Model* (Pingree Hill Publishing, Auburn, NH, 2017).

______, 2017e, *From Qubits to the Unified SuperStandard Model with Embedded SuperStrings: A Derivation* (Pingree Hill Publishing, Auburn, NH, 2017).

______, 2017f, *The Unified SuperStandard Model in Our Universe and the Megaverse: Quarks, ...* , (Pingree Hill Publishing, Auburn, NH, 2017).

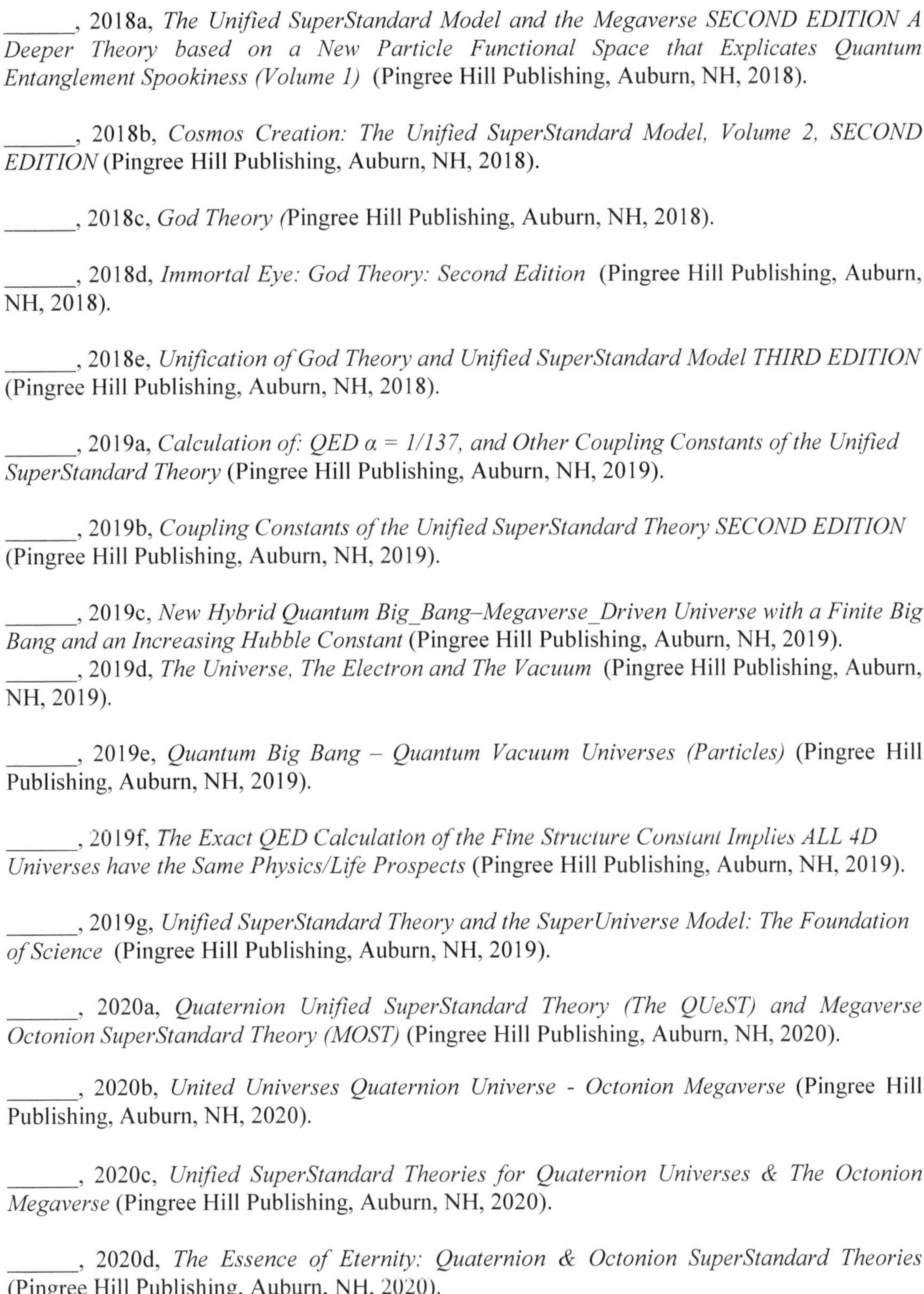

______, 2018a, *The Unified SuperStandard Model and the Megaverse SECOND EDITION A Deeper Theory based on a New Particle Functional Space that Explicates Quantum Entanglement Spookiness (Volume 1)* (Pingree Hill Publishing, Auburn, NH, 2018).

______, 2018b, *Cosmos Creation: The Unified SuperStandard Model, Volume 2, SECOND EDITION* (Pingree Hill Publishing, Auburn, NH, 2018).

______, 2018c, *God Theory (*Pingree Hill Publishing, Auburn, NH, 2018).

______, 2018d, *Immortal Eye: God Theory: Second Edition* (Pingree Hill Publishing, Auburn, NH, 2018).

______, 2018e, *Unification of God Theory and Unified SuperStandard Model THIRD EDITION* (Pingree Hill Publishing, Auburn, NH, 2018).

______, 2019a, *Calculation of: QED α = 1/137, and Other Coupling Constants of the Unified SuperStandard Theory* (Pingree Hill Publishing, Auburn, NH, 2019).

______, 2019b, *Coupling Constants of the Unified SuperStandard Theory SECOND EDITION* (Pingree Hill Publishing, Auburn, NH, 2019).

______, 2019c, *New Hybrid Quantum Big_Bang–Megaverse_Driven Universe with a Finite Big Bang and an Increasing Hubble Constant* (Pingree Hill Publishing, Auburn, NH, 2019).
______, 2019d, *The Universe, The Electron and The Vacuum* (Pingree Hill Publishing, Auburn, NH, 2019).

______, 2019e, *Quantum Big Bang – Quantum Vacuum Universes (Particles)* (Pingree Hill Publishing, Auburn, NH, 2019).

______, 2019f, *The Exact QED Calculation of the Fine Structure Constant Implies ALL 4D Universes have the Same Physics/Life Prospects* (Pingree Hill Publishing, Auburn, NH, 2019).

______, 2019g, *Unified SuperStandard Theory and the SuperUniverse Model: The Foundation of Science* (Pingree Hill Publishing, Auburn, NH, 2019).

______, 2020a, *Quaternion Unified SuperStandard Theory (The QUeST) and Megaverse Octonion SuperStandard Theory (MOST)* (Pingree Hill Publishing, Auburn, NH, 2020).

______, 2020b, *United Universes Quaternion Universe - Octonion Megaverse* (Pingree Hill Publishing, Auburn, NH, 2020).

______, 2020c, *Unified SuperStandard Theories for Quaternion Universes & The Octonion Megaverse* (Pingree Hill Publishing, Auburn, NH, 2020).

______, 2020d, *The Essence of Eternity: Quaternion & Octonion SuperStandard Theories* (Pingree Hill Publishing, Auburn, NH, 2020).

______, 2020e, *The Essence of Eternity II* (Pingree Hill Publishing, Auburn, NH, 2020).

______, 2020f, *A Very Conscious Universe* (Pingree Hill Publishing, Auburn, NH, 2020).

______, 2020g, *Hypercomplex Universe* (Pingree Hill Publishing, Auburn, NH, 2020).

______, 2020h, *Beneath the Quaternion Universe* (Pingree Hill Publishing, Auburn, NH, 2020).

______, 2020i, *Why is the Universe Real? From Quaternion & Octonion to Real Coordinates* (Pingree Hill Publishing, Auburn, NH, 2020).

______, 2020j, *The Origin of Universes: of Quaternion Unified SuperStandard Theory (QUeST); and of the Octonion Megaverse (UTMOST)* (Pingree Hill Publishing, Auburn, NH, 2020).

Eddington, A. S., 1952, *The Mathematical Theory of Relativity* (Cambridge University Press, Cambridge, U.K., 1952).

Fant, Karl M., 2005, *Logically Determined Design: Clockless System Design With NULL Convention Logic* (John Wiley and Sons, Hoboken, NJ, 2005).

Feinberg, G. and Shapiro, R., 1980, *Life Beyond Earth: The Intelligent Earthlings Guide to Life in the Universe* (William Morrow and Company, New York, 1980).

Gelfand, I. M., Fomin, S. V., Silverman, R. A. (tr), 2000, *Calculus of Variations* (Dover Publications, Mineola, NY, 2000).

Giaquinta, M., Modica, G., Souchek, J., 1998, *Cartesian Coordinates in the Calculus of Variations* Volumes I and II (Springer-Verlag, New York, 1998).

Giaquinta, M., Hildebrandt, S., 1996, *Calculus of Variations* Volumes I and II (Springer-Verlag, New York, 1996).

Gradshteyn, I. S. and Ryzhik, I. M., 1965, *Table of Integrals, Series, and Products* (Academic Press, New York, 1965).

Heitler, W., 1954, *The Quantum Theory of Radiation* (Claendon Press, Oxford, UK, 1954).

Huang, Kerson, 1992, *Quarks, Leptons & Gauge Fields 2nd Edition* (World Scientific Publishing Company, Singapore, 1992).

Jost, J., Li-Jost, X., 1998, *Calculus of Variations* (Cambridge University Press, New York, 1998).

Kaku, Michio, 1993, *Quantum Field Theory*, (Oxford University Press, New York, 1993).

Kirk, G. S. and Raven, J. E., 1962, *The Presocratic Philosophers* (Cambridge University Press, New York, 1962).

Landau, L. D. and Lifshitz, E. M., 1987, *Fluid Mechanics 2nd Edition*, (Pergamon Press, Elmsford, NY, 1987).

Misner, C. W., Thorne, K. S., and Wheeler, J. A., 1973, *Gravitation* (W. H. Freeman, New York, 1973).

Rescher, N., 1967, *The Philosophy of Leibniz* (Prentice-Hall, Englewood Cliffs, NJ, 1967).

Rieffel, Eleanor and Polak, Wolfgang, 2014, *Quantum Computing* (MIT Press, Cambridge, MA, 2014).

Riesz, Frigyes and Sz.-Nagy, Béla, 1990, *Functional Analysis* (Dover Publications, New York, 1990).

Sagan, H., 1993, *Introduction to the Calculus of Variations* (Dover Publications, Mineola, NY, 1993).

Sakurai, J. J., 1964, *Invariance Principles and Elementary Particles* (Princeton University Press, Princeton, NJ, 1964).

Streater, R. F. and Wightman, A. S., 2000, *PCT, Spin, Statistics, and All That* (Princeton University Press, Princeton, NJ 2000).

Weinberg, S., 1972, *Gravitation and Cosmology* (John Wiley and Sons, New York, 1972).

Weinberg, S., 1995, *The Quantum Theory of Fields Volume I* (Cambridge University Press, New York, 1995).

Weinberg, S., 2000, *The Quantum Theory of Fields Volume III Supersymmetry* (Cambridge University Press, New York, 2000).

Weyl, H., 1950, *Space, Time, Matter* (Dover, New York, 1950).

Weyl, H., (Tr. S. Pollard et al), 1987, *The Continuum* (Dover Publications, New York, 1987).

INDEX

About the Author

Stephen Blaha is a well-known Physicist and Man of Letters with interests in Science, Society and civilization, the Arts, and Technology. He had an Alfred P. Sloan Foundation scholarship in college. He received his Ph.D. in Physics from Rockefeller University. He has served on the faculties of several major universities. He was also a Member of the Technical Staff at Bell Laboratories, a manager at the Boston Globe Newspaper, a Director at Wang Laboratories, and President of Blaha Software Inc. and of Janus Associates Inc. (NH).

Among other achievements he was a co-discoverer of the “r potential” for heavy quark binding developing the first (and still the only demonstrable) non-Aeolian gauge theory with an “r” potential; first suggested the existence of topological structures in superfluid He-3; first proposed Yang-Mills theories would appear in condensed matter phenomena with non-scalar order parameters; first developed a grammar-based formalism for quantum computers and applied it to elementary particle theories; first developed a new form of quantum field theory without divergences (thus solving a major 60 year old problem that enabled a unified theory of the Standard Model and Quantum Gravity without divergences to be developed); first developed a formulation of complex General Relativity based on analytic continuation from real space-time; first developed a generalized non-homogeneous Robertson-Walker metric that enabled a quantum theory of the Big Bang to be developed without singularities at t = 0; first generalized Cauchy’s theorem and Gauss’ theorem to complex, curved multi-dimensional spaces; received Honorable Mention in the Gravity Research Foundation Essay Competition in 1978; first developed a physically acceptable theory of faster-than-light particles; first derived a composition of extremums method in the Calculus of Variations; first quantitatively suggested that inflationary periods in the history of the universe were not needed; first proved Gödel’s Theorem implies Nature must be quantum; provided a new alternative to the Higgs Mechanism, and Higgs particles, to generate masses; first showed how to resolve logical paradoxes including Gödel’s Undecidability Theorem by developing Operator Logic and Quantum Operator Logic; first developed a quantitative harmonic oscillator-like model of the life cycle, and interactions, of civilizations; first showed how equations describing superorganisms also apply to civilizations. A recent book shows his theory applies successfully to the past 14 years of history and to *new* archaeological data on Andean and Mayan civilizations as well as Early Anatolian and Egyptian civilizations.

He first developed an axiomatic derivation of the form of The Standard Model from geometry – space-time properties – The Unified SuperStandard Model. It unifies all the known forces of Nature. It also has a Dark Matter sector that includes a Dark ElectroWeak sector with Dark doublets and Dark gauge interactions. It uses quantum coordinates to remove infinities that crop up in most

interacting quantum field theories and additionally to remove the infinities that appear in the Big Bang and generate inflationary growth of the universe. It shows gravity has a MOND-like form without sacrificing Newton's Laws. It relates the interactions of the MOND-like sector of gravity with the r-potential of Quark Confinement. The axioms of the theory lead to the question of their origin. We suggest in the preceding edition of this book it can be attributed to an entity with God-like properties. We explore these properties in "God Theory" and show they predict that the Cosmos exists forever although individual universes (or incarnations of our universe) "come and go." Several other important results emerge from God Theory such a functionally triune God. The Unified SuperStandard Theory has many other important parts described in the Current Edition of *The Unified SuperStandard Theory* and expanded in subsequent volumes.

Blaha has had a major impact on a succession of elementary particle theories: his Ph.D. thesis (1970), and papers, showed that quantum field theory calculations to all orders in ladder approximations could not give scaling deep inelastic electron-nucleon scattering. He later showed the eigenvalue equation for the fine structure constant α in Johnson-Baker-Willey QED had a zero at $\alpha = 1$ not 1/137 by solving the Schwinger-Dyson equations to all orders in an approximation that agreed with exact results to 4th order in α thus ending interest in this theory. In 1979 at Prof. Ken Johnson's (MIT) suggestion he calculated the proton-neutron mass difference in the MIT bag model and found the result had the wrong sign reducing interest in the bag model. These results all appear in Physical Review papers. In the 2000's he repeatedly pointed out the shortcomings of SuperString theory and showed that The Standard Model's form could be derived from space-time geometry by an extension of Lorentz transformations to faster than light transformations. This deeper space-time basis greatly increases the possibility that it is part of THE fundamental theory. Recently, Blaha showed that the Weak interactions differed significantly from the Strong, electromagnetic and gravitation interactions in important respects while these interactions had similar features, and suggested that ElectroWeak theory, which is essentially a glued union of the Weak interactions and Electromagnetism, possibly modulo unknown Higgs particle features, be replaced by a unified theory of the other interactions combined with a stand-alone Weak interaction theory. Blaha also showed that, if Charmonium calculations are taken seriously, the Strong interaction coupling constant is only a factor of five larger than the electromagnetic coupling constant, and thus Strong interaction perturbation theory would make sense and yield physically meaningful results.

In graduate school (1965-71) he wrote substantial papers in elementary particles and group theory: The Inelastic E- P Structure Functions in a Gluon Model. Phys. Lett. B40:501-502,1972; Deep-Inelastic E-P Structure Functions In A Ladder Model With Spin 1/2 Nucleons, Phys.Rev. D3:510-523,1971; Continuum Contributions To The Pion Radius, Phys. Rev. 178:2167-2169,1969; Character Analysis of U(N) and SU(N), J. Math. Phys. 10, 2156 (1969); and The Calculation of the Irreducible Characters of the Symmetric Group in Terms of the

Compound Characters, (Published as Blaha's Lemma in D. E. Knuth's book: *The Art of Computer Programming Vols. 1 – 4*).

In the early 1980's Blaha was also a pioneer in the development of UNIX for financial, scientific and Internet applications: benchmarked UNIX versions showing that block size was critical for UNIX performance, developing financial modeling software, starting database benchmarking comparison studies, developing Internet-like UNIX networking (1982) and developing a hybrid shell programming technique (1982) that was a precursor to the PERL programming language. He was also the manager of the AT&T ten-year future products development database. His work helped lead to commercial UNIX on computers such as Sun Micros, IBM AIX minis, and Apple computers.

In the 1980's he pioneered the development of PC Desktop Publishing on laser printers and was nominated for three "Awards for Technical Excellence" in 1987 by PC Magazine for PC software products that he designed and developed.

Recently he has developed a theory of Megaverses – actual universes of which our universe is one – with quantum particle-like properties based on the Wheeler-DeWitt equation of Quantum Gravity. He has developed a theory of a baryonic force, which had been conjectured many years ago, and estimated the strength of the force based on discrepancies in measurements of the gravitational constant G. This force, operative in D-dimensional space, can be used to escape from our universe in "uniships" which are the equivalent of the faster-than-light starships proposed in the author's earlier books. Thus travel to other universes, as well as to other stars is possible.

Blaha also considered the complexified Wheeler-DeWitt equation and showed that its limitation to real-valued coordinates and metrics generated a Cosmological Constant in the Einstein equations.

The author has also recently written a series of books on the serious problems of the United States and their solution as well as a book on the decline of Mankind that will follow from current social and genetic trends in Mankind.

In the past twenty years Dr. Blaha has written over 80 books on a wide range of topics. Some recent major works are: *From Asynchronous Logic to The Standard Model to Superflight to the Stars*, *All the Universe!*, *SuperCivilizations: Civilizations as Superorganisms*, *America's Future: an Islamic Surge, ISIS, al Qaeda, World Epidemics, Ukraine, Russia-China Pact, US Leadership Crisis, The Rises and Falls of Man – Destiny – 3000 AD: New Support for a Superorganism MACRO-THEORY of CIVILIZATIONS From CURRENT WORLD TRENDS and NEW Peruvian, Pre-Mayan, Mayan, Anatolian, and Early Egyptian Data, with a Projection to 3000 AD,* and *Mankind in Decline: Genetic Disasters, Human-Animal Hybrids, Overpopulation, Pollution, Global Warming, Food and Water Shortages, Desertification, Poverty, Rising Violence, Genocide, Epidemics, Wars, Leadership Failure.*

He has taught approximately 4,000 students in undergraduate, graduate, and postgraduate corporate education courses primarily in major universities, and large companies and government agencies.

Recently he developed a quantum theory, The Unified SuperStandard Theory (UST), which describes elementary particles in detail without the difficulties of conventional quantum field theory. He found that the internal symmetries of this theory could be exactly derived from a 32 dimension complex quaternion theory called QUeST. He further found that a 32 dimension complex octonion theory (MOST) describes the Megaverse. It can hold QUeST universes such as our own universe. It has an internal symmetry structure which is a superset of the QUeST internal symmetries.

www.ingramcontent.com/pod-product-compliance
Lightning Source LLC
LaVergne TN
LVHW081526100826
845153LV00003B/217